D0174244

NEO-BOHEMIA

WITHDRAWN
UTSA LIBRARIES

NEO-BOHEMIA

ART AND COMMERCE IN THE POSTINDUSTRIAL CITY

RICHARD LLOYD

Routledge
Taylor & Francis Group
New York   London

THE BOHO DANCE
Words and music by Joni Mitchell © 1975 (Renewed) Crazy Crow Music
All Rights Administered by Sony/ATV Music Publishing,
8 Music Square West, Nashville, TN 37203
All Rights Reserved. Used by Permission

Lyrics for "Bohemian Like You" by Courtney Taylor-Taylor, © Dandy Warhol Music. Used by permission.

Published in 2006 by
Routledge
Taylor & Francis Group
270 Madison Avenue
New York, NY 10016

Published in Great Britain by
Routledge
Taylor & Francis Group
2 Park Square
Milton Park, Abingdon
Oxon OX14 4RN

© 2006 by Taylor & Francis Group, LLC
Routledge is an imprint of Taylor & Francis Group

Printed in the United States of America on acid-free paper
10 9 8 7 6 5 4 3 2 1

International Standard Book Number-10: 0-415-95181-X (Hardcover) 0-415-95182-8 (Softcover)
International Standard Book Number-13: 978-0-415-95181-4 (Hardcover) 978-0-415-95182-1 (Softcover)
Library of Congress Card Number 2005008803

No part of this book may be reprinted, reproduced, transmitted, or utilized in any form by any electronic, mechanical, or other means, now known or hereafter invented, including photocopying, microfilming, and recording, or in any information storage or retrieval system, without written permission from the publishers.

Trademark Notice: Product or corporate names may be trademarks or registered trademarks, and are used only for identification and explanation without intent to infringe.

Library of Congress Cataloging-in-Publication Data

Lloyd, Richard D. (Richard Douglas), 1967-
    Neo-bohemia : art and commerce in the postindustrial city / Richard D. Lloyd.
        p. cm.
    Includes bibliographical references and index.
    ISBN 0-415-95181-X (hb : alk. paper) -- ISBN 0-415-95182-8 (pb : alk. paper)
    1. Bohemianism--United States. 2. Creative ability--Economic aspects--United States. 3. Alternative lifestyles--United States. 4. City and town life--United States. 5. Wicker Park (Chicago, Ill.) 6. Artist colonies--United States--Case studies. I. Title.

HQ2044.U6L56 2005
306'.1--dc22
                                                                                2005008803

informa
Taylor & Francis Group
is the Academic Division of Informa plc.

Visit the Taylor & Francis Web site at
http://www.taylorandfrancis.com

and the Routledge Web site at
http://www.routledge-ny.com

Library
University of Texa
at San Antonio

CONTENTS

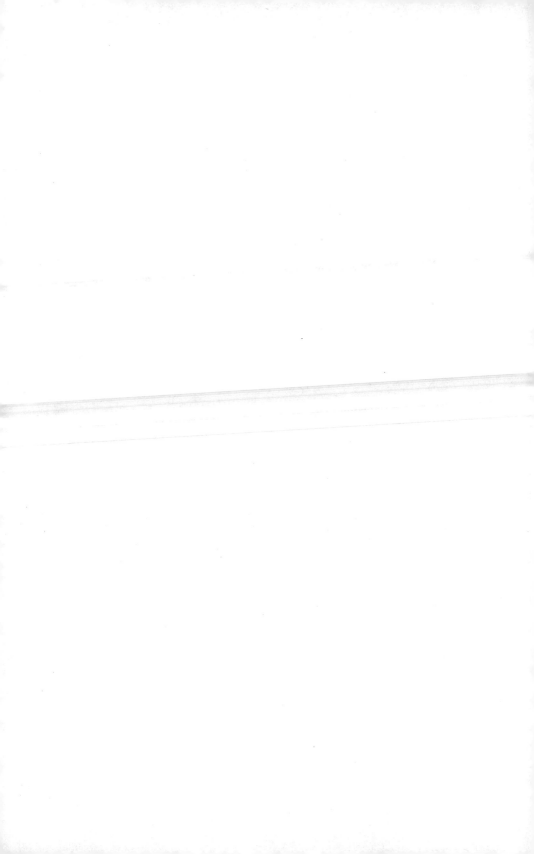

LIST OF TABLES AND FIGURES

ACKNOWLEDGMENTS

I first encountered Wicker Park in 1993, a year into my graduate work at the University of Chicago. Little did I know on that first visit (described in chapter 1) that this district would become the heart of an academic obsession that would occupy me on and off for more than a decade. The project has, at last, reached completion or, more accurately, abandonment, and I have the pleasure of looking back and acknowledging all those who were so indispensable to my life and work during that time. A longer than usual incubation period incurs a greater than usual number of personal debts. It is difficult to know where to begin.

Wicker Park was not only a site of academic observation for me, but also a place where I formed many close relationships; indeed these two aspects often merged. Sid Feldman provided a key point of entry and routine assistance in this project. I borrowed aura and more from Lucy Anderton. Important contributions also came from Jennifer Arterbury, Krystal Ashe, Bob Berger, Billy Bullion, Brad Cowley, Andy Feldman, Matt Feldman, Mike Glab, Tom Handley, Jeff Huebner, Raul Jaimez, Mark Johnson, Renee Longbons, Nina Norris, Amy Teri, Michael Weinberg, and Jennifer Zinke, as well as from the many people who sat for interviews or allowed me access into their lives. Maggie Westdale was there at the beginning, making many sacrifices for too few rewards.

At the University of Chicago, I was privileged to learn from several outstanding scholars. Many times, Moishe Postone (wisely) urged me to "take two steps back" as my enthusiasm ran ahead of my understanding, enduring my bouts of petulance with patience and grace. William Sewell is a model of the best of the academy — brilliant, innovative, and generous. Terry Nichols Clark literally rescued me from professional oblivion, finding early potential in my project on the basis of very few clues. The influence of Saskia Sassen, who directed my dissertation, began well before her arrival at Chicago and continues well after my departure. Many other past and present members of the Chicago faculty made indelible contributions, including Rob Sampson, Arjun Appadurai, Gerald Suttles, Andrew

Abbott, the late Roger Gould, Omar McRoberts, Andreas Glaeser, and Wendy Griswold. Lauren Berlant, one of the world's great close readers and critical theorists, contributed tremendous intellectual input and personal support.

As is so often the case, I learned as much from fellow students at Chicago as from the faculty. Among these, I would like to thank Anne Bartlett, Neil Brenner, Kathleen Fernicola, David Grazian, Matthew Hill, Heather Hindman, Jennifer Johnson, Jeff Link, Jeff Mantz, Jeff Morenoff, Annie Nye, Devin Pendas, Piper Purcell, Mathias Regan, Mark Schoenhals, David Smilde, James Smith, Evelyn Tenant, Rajiv Vrudhula, and Jeff Yasamoto. This list is partial at best. Monica Prasad deserves a line apart for her extensive contributions. Ezra Zuckerman read an entire late draft of this manuscript, and his advice has saved me many embarrassments. Todd Schuble of the GIS lab offered exceptionally valuable technical assistance. And also a special thanks to Thomas and Ria at Jimmy's Woodlawn Tap.

Many other scholars around the country have read parts of this manuscript as it developed, or otherwise offered support. These include Ailsa Craig, Scot Evans, Richard Florida, Herbert J. Gans, Kevin Fox Gotham, Ray Hutchison, Michael Indergaard, Jake Jacobs, Bonnie Lindstrom, Gina Neff, Leonard Nevarez, Tony Orum, Japonica Saracino-Brown, and Sharon Zukin. Krista Paulsen was especially patient and supportive during a difficult period. Harvey Molotch has gone well above and beyond the call in providing editorial advice and personal encouragement. Mitchell Duneier and Dennis Judd each read full drafts of the manuscript, contributing expceptional advice. My longtime mentor and friend Martin Sanchez-Jankowski must be singled out for the extent of his influence, dating back to my very first sociology course at the University of California at Berkeley and continuing to this day.

My colleagues at Saint Xavier University and Vanderbilt University have likewise been exceptional. Richard Fritz is a model of selfless dedication to the pedagogical vocation — I have learned much from him. Bruce Barry, Bob Barsky, George Becker, Karen Campbell, Laura Carpenter, Michael Clark, Mike Ezell, Edward Fischer, Patricia Foxen, Volney Gay, John Gutowski, Larry Isaac, Jim Lang, Richard McGregor, Sophia Sakoutis, Steven J. Tepper (the Fourth T), and Alex Trillo make coming to work fun as well as intellectually stimulating. Ronnie Steinberg and Dan Cornfield have been especially generous in supplying mentorship and encouragement. All who study the sociology of culture owe a debt to Richard Peterson. Holly McCammon, Peggy Thoits, and Larry Griffin read late drafts of chapters and made valuable comments. Gary Jensen is a model department head. Carlos Trenary has provided life-saving technical support. Linda Willingham, Pam Tichenor, Mary Griffin, and Deanne Casanova routinely rescue me from my administrative incompetence.

Students at the University of Chicago, Saint Xavier University, and Vanderbilt University have been a constant source of inspiration. I am especially grateful to the fabulous students in my seminars "The Artist and the City" and "Cities in a World Economy" at Vanderbilt University, and "Culture, Identity, and the City" at the University of Chicago. They drove me to develop much greater clarity around central issues. I am indebted to Michael Cagley, Steve Lee, George Sanders, and Damian Williams for their assistance in the research and preparation for this manuscript.

David McBride is a fabulous editor, and his efforts on behalf of this project have been extensive. Routledge's Angela Chnapko and Elise Oranges have been model professionals in helping to prepare the final manuscript.

J.J. Rosenbaum is an inspiration politically and intellectually, and was a tremendous source of support and encouragement in the last push to completion of this manuscript.

I'd also like to thank Howard Ring, Neil Kazaross, and Stan Tomchin. I owe you many debts (some very tangible): I have not forgotten.

I want to single out four friends, there from the beginning. Jeremy Straughn is a brilliant sociologist and a tremendous intellectual collaborator; even more remarkable is his generosity of spirit. Among a long list of contributions to my life, Erik Miller introduced me to Wicker Park (see chapter 1). Wilcox Snellings has seen me through many difficult times over the years, and we have shared many adventures. My old friend Steve Cronin turned me on to Jane's Addiction and the Paradise Lounge; he remains my role model for urbane sophistication.

I send all my love to my grandmother, Lillian Stratton. My aunts and uncles are pillars of support. Brian Lloyd has been an excellent brother. I am delighted to share this accomplishment with my father, Ed Lloyd. No

one deserves more love or gratitude than my mother, Karen Lloyd. I could
not be prouder of her or happier to present her with this book.

CHAPTER 1

Introduction

September 16, 1993

On a Thursday evening in September of 1993 I made my first visit to Chicago's West Side neighborhood Wicker Park. I went to see a band. I'd moved to Chicago a year earlier to begin graduate studies and found myself mostly confined during that time to the South Side's Hyde Park neighborhood. But Erik Miller, my roommate from undergraduate days in Berkeley, gave me a call to let me know that his fledgling rock band had decided to go on a self-sponsored tour of the country from its San Francisco base, and that the band would be stopping in Chicago. It was a stereotypical indie-rock odyssey, complete with the unreliable used van in which band members would often spend nights nestled among amplifiers, guitars, and Erik's drum set. They were scheduled to play at Phyllis' Musical Inn, at 1800 W. Division Street. This venue was unknown to me at the time, as was the Wicker Park neighborhood in which it was located. Still, I could hardly pass this up.

Erik has a Phi Beta Kappa key and a degree in economics, but after a year in the corporate labor force he'd decided to turn his attention to living *la vie bohème*, waiting tables and playing drums in an indie rock outfit that called itself Lost Pilgrims. This sort of thing was happening a lot in the early 1990s. The country was in recession, and many in the generational cohort that the press had taken to calling Generation X[1] — those born, like Erik and me, between 1966 and 1975 — were feeling a little disenchanted with the world of work that their unprecedented levels of education had bought them access to. As a 2002 article in *Fortune* magazine recalls, "Ten years ago grunge musicians and college-aged Cassandras who had never held a day job preached that corporate America would crush their generation's soul and leave them without a pension plan."[2] Erik had held a day job, briefly, and the experience only confirmed for him this pessimistic point of view. Of course, by the later 1990s Generation X "slackers" would be renamed "entrepreneurs" in the popular press, igniting a "new economy" with their creative, iconoclastic approaches to business. But that would be a few years off.

Lost Pilgrims got the gig at Phyllis' by cold-calling the bar. The owner, following his normal policy, said the band could play for a cut of the door charge (split with the other band in the night's lineup). Band members had to pay for their own beer, meaning that, like many of their shows, this one promised to be a more or less break-even proposition. When Erik gave me Phyllis' address, I was uneasy. I'd visited the well-known constellation of bars on Division Street in the downtown Gold Coast area, where it intersected with Rush Street.[3] But they are all about mating opportunities, not rock music or hipster culture.[4] Though Phyllis' is little more than two miles farther west on Division, my 1993 mental map of Division Street went only as far as the public housing project Cabrini Green, a vertical slum sitting at the backdoor of the high-end shopping and residential districts on the lakefront. Thus, I had no idea what to expect. Nor was my search for the bar on the big night encouraging. Today that stretch of West Division is checkered with boutiques, bars, and upscale restaurants, but back then it was pretty barren.

Phyllis' itself was an unimposing joint, and it remains so today. A corner bar, it is marked chiefly by an Old Style beer sign protruding from its façade. The sign is supposed to light up, but it doesn't. The name of the bar is painted on a sheath of wood jammed into the small front window and obscured by protective mesh wiring. This modest presentation led me to overlook Phyllis' altogether on three passes by the block. Also on the block was a vacant, rubble-strewn lot that, as I would later learn, is roughly the location of the fictional Frankie Machine's rooming-house home in Nelson Algren's midcentury novel *The Man with the Golden Arm*.[5]

Figure 1.1 Phyllis' Musical Inn. Photo by the author.

When I finally found my way to the door, I paused, uncertainly surveying Plexiglas windows that hid whatever awaited me inside.

As I would later learn, the bar has been called Phyllis' Musical Inn since 1952, when it was purchased by the Jaskot family. Back then the music was Polish old country, entertaining the white ethnic immigrants who once dominated the neighborhood population. It is named after the wife of the original proprietor and bartender, Clem Jaskot, and Phyllis herself was known to play the accordion there. Today it's run by Clem Jr., who can be found behind the bar in the early evenings. Its interior is tiny, and uncomfortable wooden auditorium seats are bolted into its floor facing the stage. A small space is left open for audience members who want to dance or slam into each other during the performances. Around the stage is a colorful mural depicting the Chicago skyline with a piano keyboard snaking through it, painted in the early 1990s by the neighborhood artist Shelley Hutchison. The mural extends all around the wall, and tucked into it is a depiction of a merry Phyllis on the accordion, and Clem Sr. laughing behind the bar.

Figure 1.2 The Stage at Phyllis' Musical Inn. Photo by the author.

The music has changed considerably since 1952, having made the switch to pounding rock 'n' roll in the 1980s. When I walked through the door, I was confronted with a surprisingly thick crowd of young people. These patrons were a striking bunch, attired in funky secondhand ensembles and sporting a variety of tattoos and piercings. This was a side of Chicago that I had not yet seen. Shortly afterward, I would learn that Phyllis' Musical Inn was in fact part of a Wicker Park scene attracting significant attention as a seedbed of cool X-generational culture. Indeed, the music industry's trade standard *Billboard* magazine had published a map of the neighborhood only weeks earlier,[6] marking little Phyllis' Musical Inn among the attractions.

Lost Pilgrims was the first of two bands on the bill, and it was the second act that attracted the unusually large Thursday night crowd. It was a local outfit called Veruca Salt (named after the spoiled rich girl in *Charlie and the Chocolate Factory*), making one of its earliest public appearances. Though scarcely exposed, the neighborhood buzz around Veruca Salt was strong, traveling through a thick local network of new culture hawks.

I would learn in subsequent years that one had to be braced for considerable disappointment when sampling new fare at places like Phyllis', but that night Veruca Salt delivered the goods, delighting the crowd with a thundering set that fairly shook the little bar. Barely a year later, Geffen Records released the band's debut album, and Veruca Salt shot into the rotations on alternative radio stations and on MTV with its hit song "Seether." In what proved to be a common pattern, this rise to fame resulted in Veruca Salt's repudiation by its former neighborhood supporters.[7] But that night in 1993, there was unfettered enthusiasm from all of us, experiencing the cutting edge there in that obscure dive.

I was hooked.

April 24, 2000
(Adapted from field notes)

It is a Friday night around 10 o'clock on Wicker Park's North Avenue, at the opposite end of the neighborhood from Phyllis' Musical Inn. As usual, the intersection around North, Damen, and Milwaukee Avenues is jammed with young people scurrying to avail themselves of the hip local nightlife. By now, well into an ethnographic investigation of Wicker Park's local culture and economy, I am about to make my first contact with an exemplar of the most recent new sector of neighborhood commerce, Buzzbait, an Internet design boutique on the sixth floor of the neighborhood's Northwest Tower. Barely past a year old, Buzzbait has expanded rapidly and is having a party to signal its collective sense of well-being and optimism. Incidentally, Erik Miller, my drum-playing ex-roommate, is with me — he's out to visit from San Francisco. Erik still plays in the band, now called the Keeners, but like many a former slacker musician in San Francisco, his day job is with a dot-com start-up.

Outside the Northwest Tower, which is nicknamed the Coyote and which lends its name to the neighborhood's annual arts fair, I run into the friend who tipped me to the event in the first place. This shaven-headed neighborhood insider and aspiring screenwriter is at that moment "waiting for the man," as the Velvet Underground song puts it — that is, a drug dealer will be coming by soon for a prearranged meeting, and they will transact. This sort of commerce is no stranger to the corner; indeed, not much more than ten years earlier, it was pretty much the only commerce that was there. "It's not for me," my friend tells me. "This girl inside asked me to get four grams [of cocaine]. And I hate doing this, but she's wearing a very short skirt." Erik and I leave him on the corner and go into the party.

The Northwest Tower has an old elevator with retractable cage doors and a real-life grizzled old man as its operator. I am told he doesn't talk much, but everyone in the building loves the old man. The door opens onto the sixth floor, and we are deposited directly into Buzzbait's offices, where the party is in full swing. For the most part, the crowd looks an awful lot like a crowd I might have seen at any neighborhood party in an artist's loft over the last several years. Attire is thrift-store chic. Michael Weinberg, one of Buzzbait's founders, greets us. He's also a musician who likes to throw jam parties at his loft. I ask him where the drum set is. "I keep 'em at home. I don't live here yet. Close, but not yet." As in Wicker Park's neighborhood bars, the work of local painters dots the walls. There is plenty of such work to go around; during the 1990s, Wicker Park was considered among the most vibrant arts communities in the country, and the neighborhood Flat Iron Building alone provides studio space shared by more than one hundred artists.[8]

I spot Big Mike Glab, a contributing writer for the local alternative weekly *Chicago Reader*. Big Mike has recently been subcontracted by Buzzbait to do some copywriting. In his early forties, this longtime denizen of the Chicago arts scene is looking especially sharp this evening, as I tell him. "You should see the sharkskin jacket I have in the closet," Big Mike responds. A few weeks earlier he had published a *Reader* cover story profiling the extreme performance artist William Dark and his traveling freak show. Tonight Dark is at the party, and he's currently entertaining the crowd of tech workers and hipsters by shoving a wire hanger through his nose. Big Mike brokered the gig. "That's disturbing," I tell Big Mike. "Oh, it's disturbing," he replies happily.

This is a decidedly noncorporate office party, appropriate to the hipster ethos that attaches itself to the Internet business culture claiming to transform the world. Around the party, rumors circulate about impending buyouts and IPOs, and the fantastic jackpots that have already been paid

Figure 1.3 William Dark Entertains Buzzbait. Photo courtesy of Michael Weinberg.

out to firms not unlike this one in districts such as San Francisco's South of Market and New York's Silicon Alley. Silicon Alley's Razorfish is the ur-example, and it's Weinberg's inspiration.[9] My screenwriter friend, having completed his mission out front, joins Erik and me and whispers that a $5 million offer had already been made to buy Buzzbait by an unnamed source. These tidbits are always whispered, as if saying them too loudly will undo the fantasy that one can be bohemian and get rich too.

I find myself talking with a young woman in red leather pants. A classically trained musician, she tells me about the CD she just finished recording at her own expense, describing it as "experimental, avant-garde classical music." After a while she asks me about my own work, and, no covert ethnographer, I tell her that I am doing a sociological study of the neighborhood. "Oh," she says, "gentrification." Then she narrows her eyes: "Are you for it or against it?" It turns out that she is profoundly opposed to the escalation of investment on Chicago's West Side. Oddly, she does not seem to think that attending a party for a hot young Internet boutique in the heart of the West Side gentrification scene is inconsistent with those politics.

Chicago's "Burst of Bohemia"

As these anecdotes suggest, the 1990s were an eventful decade for the
Wicker Park neighborhood, during which it transformed from a relatively
obscure and depopulated barrio into a celebrated center of hip urban
culture. Locally produced paintings adorned the walls of neighborhood
galleries and also of its local bars, coffee shops, and even Internet firms,
and the music of homegrown performers, as well as guests like Lost
Pilgrims and later the Rolling Stones, drifted from smoky clubs to the
nocturnal sidewalks. Having honed their talents in intimate venues like
Phyllis' Musical Inn, Veruca Salt and a handful of others found a wider
audience (though aspirants exceeded opportunities by a wide margin),
exported from the neighborhood by the octopi of corporate culture industries, which in the 1990s disguised some of their tentacles as pseudo-
"indie" labels, [10] and by the much-vaunted avatar of youth culture, MTV.
Similarly, the neighborhood was a staging ground for young artists in
other popular and avant-garde arts, including theater, film, sculpture,
and painting, some of whom also exported their products and talents into
more established, higher stakes culture markets.

In addition to feeding new products into the flexible networks of cultural production, Wicker Park also served as a strategic site for other local
and extralocal economic interests that contributed an infusion of capital
into the once-moribund neighborhood economy. A host of new entertainment venues opened, many themed in accordance with the local ethos of
funky, hip, and creative culture, from galleries and performance venues to
bars, boutiques, restaurants, and coffee shops. Educated young professionals, arguably attracted to the cachet of the newly hip neighborhood address
as well as to these emergent local amenities, moved into the neighborhood
in increasing numbers, leading to a dramatic escalation of home values and
rents. Alongside the arts and entertainment enterprises were a number of
small graphic design and Internet design firms like Buzzbait, advertising
their creative credentials with a Wicker Park address and often employing
local artists as creative labor. Indeed, before my eyes Wicker Park unfolded
as a serendipitous site containing to greater or lesser degrees a host of
economic and cultural trends of concern to theorists of postindustrial

urbanism and postmodern culture. It was even, in 2001, the stage for the filming of a reality television program.

As I would discover, such activity stood in stark contrast to decades of decline that had preceded it on Chicago's West Side, and it provided little benefit for the mostly Latino residents who had endured the associated hardship. Wicker Park had once been a bustling site of industrial enterprise, mixing low-rise factories and warehouses with the residences of immigrant laborers. Though most of these lived a meager existence, the neighborhood also contained pockets of prosperity that account for the presence of the comparatively fine old homes ringing the park, a few blocks from the Northwest Tower.[11] By the 1980s, those days were long gone, and Wicker Park visibly bore the scars of deindustrialization, deterioration, and population decline. The white ethnic, primarily Polish, population that lent the neighborhood much of its distinctive cultural identity in the first half of the twentieth century had largely given way to the city's newer immigrant groups, Mexicans and Puerto Ricans. These more recent arrivals encountered a landscape of much more tightly circumscribed opportunity than their predecessors had.

During the 1980s, few outside the neighborhood seemed particularly interested in Wicker Park. Media coverage was limited to the occasional lurid crime story, along with polite notice of the annual neighborhood Greening Festival, a harbinger of the much higher profile arts festivals to come.[12] At the tail end of the decade, there was some recognition that an art gallery or two had opened,[13] but the neighborhood was not widely considered exceptional. Even as an exemplar of urban decay, it was decidedly second-rate by Chicago standards, unable to compete with the exceptionally gory details of life on the hypersegregated South Side.[14]

While Wicker Park was ignored in the media, local artists were already taking increasing interest in the area during the 1980s, and its reputation spread through those informal channels maintained by people concerned to find the cutting edge. For the "starving" young artist, the neighborhood offered a range of material advantages, principally inexpensive rents and the kinds of derelict industrial spaces that could be reconfigured as studios and performance spaces. Moreover, the neighborhood is strategically located, proximate to the downtown Loop and to the near North Side's sprawling network of independent theaters and musical performance

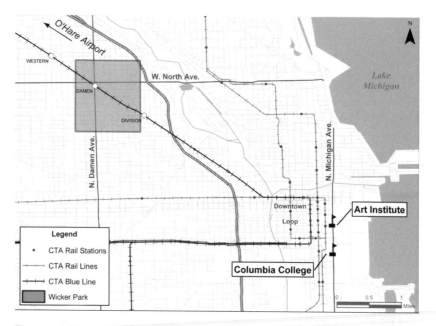

Figure 1.4 Map of Wicker Park and Chicago, 2001. Map created by Richard Lloyd and Todd Schuble.

venues. The Elevated Train's Blue Line bisects the neighborhood, with the Damen Avenue stop depositing travelers near the hub of West Side activity at the six-cornered intersection of North, Damen, and Milwaukee. The line tethers Wicker Park to the Loop, and in the other direction, to O'Hare International Airport; just as important, it stops within a block of the School of the Art Institute in the South Loop (see figure 1.4). Current and former students at that institution and at the proximate Columbia College were important denizens of the developing arts scene.

On the other hand, it lacked some important amenities also, and this lack meant that although artists had trickled in at least as early as the mid-1980s, their ability to constitute a genuine "scene" such as the one that would emerge in the next decade remained only incipient. In the course of this book, I will examine how the slow achievement of a critical mass of artists and affiliated "countercultural" sorts, coupled with other local and extralocal developments, created the context for the sudden explosion of an artistic scene that garnered much wider attention.

By 1993 — around the time that I wandered uninitiated into the neighborhood — the word was getting out. Individuals with aspirations in a range of creative pursuits, from the rarefied fine arts to more popular genres such as music and film, were crowding into local lofts and flats, but it was the music scene that initially struck the most forceful chord with the wider public. Anointed the "new capital of cutting edge" by *Billboard* magazine,[15] Wicker Park was descended upon by industry gatekeepers who snapped up such local acts as Veruca Salt, Urge Overkill, and Liz Phair from among an ever-growing legion of aspirants. This flurry of attention shone a light on the neighborhood as a diverse package of cutting-edge innovation, and multiple interests took note. Wicker Park's stay atop the putative throne of hip new musical production was not long — with those on the cutting edge, public announcement typically also amounts to an obituary.[16] Media purveyors of the hip turned their attention elsewhere fairly quickly, but by that time the neighborhood ethos was constituted not only in headlines or even by the massing of creative aspirants, but also by local institutions with greater or lesser staying power: the local galleries, coffee shops, and bars that continued to announce, and trade upon, bohemian notoriety.

Thus, though media record of the fact settled quickly after 1994 to a less hysterical pitch, Wicker Park was still known as an artists' neighborhood, and that has remained a durable perception throughout neighborhood changes that included intense escalation of local rents. This identity is legible in large part due to the perceived continuity with past congregations of artists in cities. Indeed, even the decidedly un-hip national newspaper *USA Today* declared Wicker Park to be "the Windy City's burst of bohemia" in 1995,[17] and in 2002 *The New York Times* travel section celebrated it as Chicago's bohemian hub of funkiness and creativity.[18]

Wicker Park is understood, both by media observers and by local insiders, in the context of the mythic tradition of bohemia, shorthand for both a distinctive sort of urban district and an associated style of life. As we will see, this is a construct deeply embedded in the history of the modernist metropolis and of modernist cultural innovation. The origins of the term as a designation for the artists' quarter lie in nineteenth-century Paris, tagged by Walter Benjamin as "the capital of the nineteenth century," for him the very cradle of aesthetic modernism.[19]

The reader may have already surmised that I take this continuity seriously, since I have devised the term "neo-bohemia" as a heuristic to examine the changes that took place in Wicker Park during the 1990s and beyond. Bohemia is more than just a likely cliché deployed by media observers. In fact, the traditions of the artist in the city, shaped both by material exigencies and cultural identifications, create a blueprint for contemporary action in a neighborhood like Wicker Park. In part this is because local creative aspirants in Wicker Park face many challenges similar to those encountered by their bohemian predecessors. But it is also because even where a given participant may resist identifying with the *term* bohemia — and some do, vehemently — their ideas about what constitutes an appropriate artists' lifestyle remain profoundly influenced by its legacy. Indeed, even resistance to the term reveals its powerful resonance with the neighborhood scene and its participants.

In a 1993 *New York Times* feature article titled "Edgy in Chicago: The Music World Discovers Wicker Park," the local musician Tim Rutilli is quoted as insisting, "It isn't Paris in the 1920's. It really isn't."[20] The comment is not accompanied by any further elaboration. The connotations of Paris in the 1920s are presumably well understood by Rutilli, by the *Times* reporter, and, just as important, by the typically well-educated *New York Times* reader. By "Paris in the 1920's," we assume that Rutilli is signaling the felicitous congregation of modern artists, including American expat writers like Hemingway, Fitzgerald, and Stein, living the bohemian life in Parisian cafés.[21] While recent observers like Richard Florida report that contemporary creative sorts reject the label of bohemian because it carries too much negative baggage,[22] Rutilli does not seem to be repudiating the value of the Parisian model, but rather to be modestly acknowledging that the intensity of creative activity and romantic striving in Wicker Park could not match the standard set by its Parisian precursor. But even in the negative, the comparison demonstrates that both neighborhood insiders and interested observers were likely to interpret the local artists' scene in Wicker Park as somewhere on a continuum with highly recognizable and richly connotative past bohemian instantiations. Would the quote have sounded much different if Rutilli had substituted 1950s Greenwich Village or *fin de siècle* Vienna?

Certainly, though, Rutilli is correct, and not only because the cream of Wicker Park's 1990s artists are unlikely to find their way into the canon with the same force or frequency as did the most successful artists of 1920s Paris. Richard M. Daly, Chicago's mayor in the 1990s, has made clear his admiration for Paris, pining to lend Chicago some of its aesthetic appeal,[23] but Chicago is still not Paris, and the last decade of the twentieth century was very different from the third. Despite claims to subcultural autonomy, it is reasonable to assume that the activities of artists unfold in the context not only of cultural tradition, but also of the distinctive metropolitan dynamics in which they are embedded. For obvious reasons, 1920s Paris could not magnetize recording-industry scouts and digital designers as Wicker Park did. Moreover, though the artists working in 1920s Paris may have made a central contribution to the modernist canon, their effect on the Parisian economy of the time was likely far more negligible. Thus, while Wicker Park's emergence as a center of hip culture bears enough similarity to Paris' in the 1920s for Rutilli's comment to make sense, it is also distinguished from this predecessor by its encounter with a distinctive — and dynamic — urban landscape.

Postindustrial Chicago and the New Bohemia

Belying the apparent inertia of concrete and steel, cities are dynamic entities, and the late twentieth century has witnessed dramatic transformation in the structural foundations of urban development. Current debates on the contemporary Western metropolis focus on the changing economic forces that drive urban growth and decline; these include deindustrialization, globalization, and associated increases in the importance of immaterial labor in areas like finance, technology, and media design.[24] The vaunted "LA school" of urban studies stresses patterns of deconcentration that seem to obviate the dense neighborhood morphology in which traditional bohemias have been located.[25] Others indicate that alongside sprawling growth, we also find the resurgence of older downtowns and select inner-city neighborhoods.[26] This neighborhood-level development, unevenly distributed throughout older cities, no longer follows the impetus of industrial production. Instead, new patterns of production characterize the city and its neighborhoods, with a larger role for culture and technology. It is in this context that a new bohemia like Wicker Park emerges, and we will focus on the new role that the artists' neighborhood plays under these circumstances.

Though Chicago has been perhaps the most studied city in the world, using it to examine either postindustrial trends (other than those toward decay and dissolution) or cultural innovation may seem at first glance unlikely. The association between Chicago and the period of immigrant- and industry-fueled urban growth that characterized U.S. cities in the first half of the twentieth century is profoundly inscribed in the traditions of urban studies. Rough-hewn blue-collar workers, not eccentric aesthetes, typically are taken to be the exemplars of the city of broad shoulders. University of Chicago sociologists occupied a hegemonic position in the field during the industrial period, constituting their home base both as local laboratory and, implicitly, as the model *par excellence* for the modern metropolis.[27] As the second half of the century brought increasing changes in urban economy and morphology, the effort to capture new dynamics seemed to necessitate a repudiation of the older Chicago School, with new prototypes of a postmodern urbanism duly nominated. Michael Dear, in particular, has repeatedly made explicit his view that Los Angeles has replaced Chicago as the proper focus for the study of the "postmodern urban condition."[28]

In fact, while Chicago has slipped to third in the population rankings of U.S. metro areas, the city has nonetheless weathered the transition to the postindustrial period, albeit in an often painful and highly uneven fashion. Along its South Shore, the apocalyptic consequences of deindustrialization are painfully evident as impoverished residents, overwhelmingly African American, occupy dilapidated residences alongside the deteriorating relics of industrial enterprise. But in the central business district — the famed Chicago Loop — postindustrial enterprises in sectors such as finance and advertising thrive.[29] Moreover, in Wicker Park, the pattern of postindustrial decay that set in during the immediate postwar decades has given way to a new style of redevelopment, combining its celebrated arts community with residential gentrification by white-collar professionals, the injection of new media and design enterprises, and a thriving entertainment economy.

I was a witness to much of this change, first as an interested consumer of the local youth culture and later as a social researcher. I was interested in establishing on-the-ground shifts in the social, economic, and demographic makeup of the neighborhood as they related to the creative community. As I delved into the local transformation of Wicker Park, I came to understand that these developments could not be properly interpreted as a reversal of prior trends in the neighborhood but rather a redirection — a dramatic transformation in the structural foundations of neighborhood social and economic activity.

Moreover, I became convinced that these changes could not be accounted for only by examining the activities of local participants, either artists or entrepreneurial capitalists. Pierre Bourdieu makes the trenchant observation, "The perfectly commendable wish to go see things in person, close up, sometimes leads people to search for the explanatory principles of observed realities where they are not to be found (not all of them, in any case), namely at the site of observation itself."[30]

Sited observation, employing on multiple methodological strategies, provides my direct empirical contribution, but it is only one half of the project undertaken. In order to make sense of the local dynamics observable in Wicker Park, we need to understand the larger transformation of Chicago's economy and demography: the decline of the industrial base that had once sustained neighborhoods like Wicker Park; the ascendancy of postindustrial enterprises, led in Chicago by advertising, finance, and technology; and the growing importance of a new class of educated professionals servicing these industries. Grasping these dynamics, in turn, required expanding my conceptual template to incorporate shifts in the nature of global capitalism.

There is a symbiosis within a new bohemia like Wicker Park between the activities of contemporary artists (and affiliated lifestyle aesthetes) and central features of postindustrial urbanism. In fact, this notion — that the activities of artists provide benefits to local economies far beyond what can be measured by formal art markets (including large-scale popular culture industries) — is becoming less controversial as the measures supporting it multiply. Many have noted the role that artists play as the vanguard of a distinctive sort of gentrification, in which underused spaces inherited from a city's industrial past are reconfigured as live/work spaces and galleries. Such processes have been especially well studied in New York, the city with the richest artistic tradition in the United States, and indeed arguably the capital of modern art in the Western world, at least since the Second World War.[31] Sharon Zukin's work in SoHo during the 1970s is particularly exemplary, and her notion of "the artistic mode of production" anticipates trends that are foundational to this analysis.[32] More recently, gentrification in the East Village has been the subject of several fine studies.[33]

By the late 1990s, scholars working both with quantitative and qualitative techniques were beginning to note the relationship between the concentration of artists in a given district or region and the presence of other crucial postindustrial enterprises, principally enterprises in the "new economy" of digitally based enterprises. Comparative work among metro areas shows that relatively high proportions of artists within a region correlate to increased concentrations of high-tech enterprise and enhanced levels of overall economic development.[34] To these quantitative projects, the extended case analysis of Wicker Park adds local depth and nuance, opening up new avenues of explanation.

If, however, we insist on understanding the neighborhood's transition in the context of the past practical activities and mythic imagery of bohemia, the much-noted connection between artistic concentration and capitalist strategies presents itself as surprising and paradoxical. After all, the ideology of bohemian life has been constituted as oppositional to the propertied classes ever since it arose in nineteenth century Paris.[35] Then that meant the repudiation of the bourgeoisie by bohemian artists. As we will see, this remains a potent part of the local belief system in Wicker Park, though bourgeois shopkeepers have been replaced by soulless yuppies as the principal foil. Moreover, in Wicker Park today, animus is also directed toward the presumed dictates of a more faceless foe — *the corporation,* a specter whose shape in the contemporary bohemian imaginary I will argue actually derives from the distinctive forms of bureaucratized capitalism of the mid-twentieth-century United States. In any case, bohemia has long been assumed to be antithetical to the instrumental, rationalized strategies of capitalism, an assumption shared by its adherents. At the very least this has meant that past bohemian congregations were marginal aspects of an urban economy. For critics like Daniel Bell, they were considered genuinely threatening to the economic order.[36]

Indeed, this traditional framing underscores Richard Florida's assertion that while artists matter a great deal in today's economy (though not, it turns out, primarily as producers of art), the term "bohemia" is obsolete because artists no longer feel themselves to be alienated within the current economic order.[37] While indebted to his empirical contributions, I am not convinced by Florida's claim for bohemia's conceptual obsolescence, and I therefore advance two propositions that retain bohemia as a central but conceptually reconstructed category. The first is that there is significant continuity between the artistic congregation in Wicker Park and past districts in the modernist metropolis that have worn the mantle of bohemia; indeed, that the cumulative imagery of the artist in the city is directly inscribed on the practical strategies of contemporary social actors. The second suggests that even as this is true, the new bohemia of the late twentieth and early twenty-first centuries plays a necessarily novel role in enhancing the interests of postindustrial capitalist enterprises, especially property speculation of various sorts, entertainment provision, and new media production. These propositions will be tested against my observations in Wicker Park, as well as against the evidence contained within the scholarly, artistic, and popular record (since they are historical-comparative, and cannot be addressed via the site alone). Only then can we begin to arrive at a solution to the paradox of how a sociocultural space that has long been understood in terms of marginality within and in opposition to the capitalist economy can suddenly emerge as a source of comparative advantage for new capitalist strategies.

This book is divided into three parts. Part I, "Industry and Art in the Modern City," supplies theoretical and historical background essential to the subsequent analyses, addressing the local history of Wicker Park in the context of broader social changes, and the historical genesis of the sociocultural patterns of bohemia from the Parisian prototype to the postindustrial present. Part II, "A Postindustrial Bohemia," traces redevelopment in contemporary Wicker Park. We will examine how the artists' scene came together in the late 1980s and early 1990s, in which a landscape of postindustrial decay was increasingly interpreted as edgy and glamorous, laying the groundwork for new styles of economic development. As the 1990s wore on, the neighborhood experienced increasing residential gentrification and growing celebrity as a center of cultural innovation and hip entertainments. These developments are analyzed as a consequence of the encounter between the traditions of the modernist bohemia and the demographic and economic changes of postmodern urbanism. Part III, "Artists as Useful Labor," shows how the dispositions and competences nurtured in the new bohemia contribute to patterns of culture industry production and other styles of postindustrial production, particularly local entertainment provision and new media enterprises. The conclusion, "The Bohemian Ethic and the Spirit of Flexibility," ties these strands together in order to make sense of the role that the new bohemia plays in the context of flexible, global capitalism.

PART I

Industry and Art in the Modern City

CHAPTER 2

Production and Neighborhood

Every epoch, in fact, not only dreams the one to follow,
but, in dreaming, precipitates its awakening.
— Walter Benjamin, "Paris, the Capital of the
Nineteenth Century," 1935

THE SQUAT BRICK STRUCTURE on North Avenue in Wicker
Park does not appear remarkable, especially when compared
to the ornate façade of the Flat Iron Building across the street.
In the early part of the twentieth century, the brick building
housed a dressmaker's sweatshop and its cavernous interior
was crammed with immigrant women working for piece rates.
By the 1970s, the building had fallen into disrepair, languish-
ing as a storage facility, while the neighborhood's population
was progressively eroding. In the 1980s, it housed a "shooting
gallery" where heroin and other narcotics were sold and consumed. Outside, street prostitution thrived. Such seedy commerce might be read as the consequence of an economic shift
that could only leave this building a relic, a sign of the ongoing
displacement of industry that robbed the neighborhood of a
more legitimate economic base.

But in 1989 the old brick building became home to a new
occupant, the Urbus Orbis Café. Urbus Orbis provided a key
amenity for what was at that point a nascent artists' scene in the
surrounding neighborhood, serving as a hangout, display venue,
and platform for collaboration. The opening of Urbus Orbis

marked a new turn in the neighborhood's identity, and in its modest lifetime
the café was hailed as a premier site in Chicago's new bohemia, where artists,
musicians, and young professionals could sip coffee and admire the locally
produced artwork decorating Urbus Orbis's brick walls.

Despite its popularity, by 1998 Urbus Orbis had succumbed to the
many perils that beset small businesses, including the advancing gentrification of the neighborhood, which made operation prohibitively expensive.[1]
An antique store followed with an even shorter tenancy in the same space.
But for a memorable interlude in 2001, a tenant with considerably deeper
pockets rehabbed the building into an odd combination of television studio
and dormitory. It became the home for an installment of MTV's popular program "The Real World," a pioneer of the reality television genre.
"The Real World" sets up an eclectic cast of young people in a domicile
of putative urban cool and turns cameras on their presumably unscripted
experience over a period of months, editing the outcome into a season of
episodes for the vicarious entertainment of a global audience. Prior locations include New York City, San Francisco, Seattle, and New Orleans;
a key part of the story told is the inherent glamour and drama of young
people participating in the hip consumerism offered by the big city.

Its tenancy in Wicker Park marked the show's first foray into Chicago,
and MTV's selection of this loft to stage its exercise in television semi-vérité
ratified the ongoing status of the neighborhood as the city's hipster ground
zero, a designation that Urbus Orbis had helped establish. But if "hip" was
once constituted in connection with outlaws and alienation,[2] it is now an
urban amenity, exportable through the circuitry of the mass media and
fashion industries. Meanwhile, local artists and hip kids were markedly
ambivalent toward MTV's co-optation of a neighborhood aesthetic over
which they feel proprietary. This tension was nothing new. As I discovered
during my years in Wicker Park, the contemporary bohemia constituted
there was routinely mined for profit, often against the opposition of those
who "make the scene." Moreover, while the attraction of artists to distinctive
urban districts has a long history, the opportunities to exploit the bohemian
ethos multiply in our media- and culture-saturated present.[3]

The strange odyssey of this squat brick structure — from light industry
to residential media object — is a good place to begin an examination of

Wicker Park's role in the changing geography of capitalist accumulation. Neighborhood life in Chicago once was structured by industrial manufacturing; steel and stockyards used to be the story, but it is now culture and technology. Correspondingly, the building has shifted from sweatshop to postmodern distraction factory, where everyday life, leisure, and image production merge. Trendy nightclubs, artists' lofts, and the offices of multimedia design firms occupy similar structures around the neighborhood. Local residences that once housed a blue-collar labor force now accommodate artists, students, and educated young professionals who thrive on the local ambiance of urban cool.

The dynamic shifts in local economy and demography are embedded in social processes that are global in scope, including the displacement of manufacturing from old urban centers to suburban locales and, later, to sites outside the United States. Left behind are both the workers who had once sustained stable inner-city blue-collar communities on manufacturing paychecks, and the outmoded fixed capital of light industry. Meanwhile the increased mobility of capital, allowing a new global division of labor, has been accompanied by the heightened aestheticization of the economy. As Fredric Jameson notes of this postmodern period, "What has happened is that aesthetic production today has become integrated into commodity production generally: the frantic economic urgency of producing fresh waves of ever more novel-seeming goods . . . at ever greater rates of turnover, now assigns an increasingly essential structural function and position to aesthetic innovation and experimentation."[4]

Historically, Wicker Park belongs to the mode of standardized industrial production that characterized Chicago's explosive growth in the first half of the twentieth century, with poorly educated immigrants performing routinized tasks. Such production seems far removed from the aestheticized economy that Jameson describes. The factory was fixed, solid, local, whereas the postmodern economy is often characterized as liberated from spatial constraints.[5] However, recent work by Harvey Molotch indicates the profound role that distinct places have played, and continue to play, in the seemingly ephemeral activities associated with design and aesthetic innovation.[6] Moreover, these activities are still carried out by people, a growing class of culture workers. Thus, in its late-twentieth- and early twenty-first-century incarnation, Wicker Park organized a labor force largely directed toward the aesthetic innovation and experimentation whose "essential structural function" Jameson identifies. Rather than being merely anachronistic, neighborhoods like Wicker Park, once predicated on the spatial practices of blue-collar manufacturing, are reconfigured as strategic sites in the aesthetic economy. But before we can tell that story, we must first turn our attention to the historical processes within which local space is constituted and reconstituted.

Spatial Practices

The example of this brick building provides a lens into the dynamism that characterizes urban environments, a dynamism not limited in Wicker Park to the spectacular transitions at the end of the twentieth century. As a simple configuration of brick and mortar, there is nothing exceptional about the building. Such structures proliferate throughout the city of Chicago, the artifacts of the city's industrial past. Yet these buildings need not be simply relics, the manifestation of "dead" labor and bygone economic strategies. They continue to be inscribed by what Henri Lefebvre calls "spatial practices,"[7] the practical activities that shape and reshape material space, and which include both active rehabilitation and passive neglect.

Quite obviously, few such structures have followed as colorful a trajectory as did the Urbus Orbis/*Real World* building — after all, how many

Figure 2.1 The Brick Building. Photo from ChicagoFun.org.

buildings get to star on MTV? Some have been rehabilitated to new uses — living lofts, office spaces, or entertainment venues, for example. Others remain dedicated to surviving strands of light industry in the city's shriveled industrial economy. Still others are abandoned to decay, a fate our building once seemed destined to share.

While any of these cases must be understood partly in the context of historical contingency and distinctive locality, we must also take into account broader cultural and economic forces. The economic activities animating the building in the late twentieth century are part of a complicated and dispersed economic system that also incorporates downtown office towers and Third World sweatshops. The production of images may take on expanded importance in this global economy, but other sorts of production are not so much obviated as outsourced. Moreover, the suddenly marketable image of the building — as ground zero within a neighborhood known for creative energy and hipster diversions — makes sense in the context of historically emergent meanings attached to certain sorts of urban districts.[8] Our particular building is thus exceptional, but it is also emblematic. This is because both the building and the neighborhood surrounding it point to processes beyond themselves, and we will see how in these exceptional cases something of the contemporary general situation is thrown into bold relief.[9]

Changes in capitalist organization are evident not only in the products produced or in employment relations, but also in the social production of urban space. Indeed, across a wide spectrum of urban scholarship there is a consensus that the economic foundations of metropolitan viability have shifted in recent decades, even if the nature of this shift is contested. At the same time, cities remain historically cumulative projects, and new strategies within urban space are shaped by the interaction with prior outcomes. The dynamic encounter between the old and the new is especially evident in a city like Chicago, whose built environment is so profoundly influenced by a mode of economic organization that most experts would agree has been superseded: that is, a national, urbanized economy geared toward mass standardized production. Thus, while arguing for Chicago's postindustrial relevance as a "global city" (at least in some respects), Saskia Sassen concedes that "much of Chicago's built environment responds to earlier locational logics and its distinct construction and transport options."[10] These apparently obsolete structures pose a problem for those interested in exploiting the city for new avenues to profit. Writes Rachel Weber, "The accumulation process experiences uncomfortable friction when capital, (i.e. 'value in motion') is trapped in steel beams and concrete. . . . Prior investments create path dependencies that, because of the difficulties inherent in modifying physical structures, constrain future investments."[11] What our building suggests, however, is that these structures need not always be sources of constraint. They can also be opportunities, even under quite different social and economic circumstances than those that gave them birth. This has proved to be true not only of the building in question, but also of its surrounding district.

As a sweatshop, the building required the proximity of immigrant laborers for performing routine tasks. It also reflected the market and political forces associated with the nascent stages of large-scale industrialization. By midcentury, neither condition prevailed. Mass European immigration had been choked off after the Second World War — an at least temporary interruption in the stream of exploitable workers.[12] Other features of the mature mass-production economy likewise undermined the sweatshop model of production, including the growth of labor unions and the expanded intervention of the state on behalf of workers' rights.[13]

Thus, in Chicago, buildings like this one were already becoming obsolete for manufacturing even as factory production flourished elsewhere in the city.[14] By the 1970s and 1980s, Latino immigrants had replaced Poles as the dominant local population, and they encountered not only restricted labor opportunities as manufacturing declined, but also the increasing abdication of the state as a service provider for disadvantaged populations. As countless observers have noted, the spaces left vacant by the flight of industrial capital in cities around the United States have been filled in part by street entrepreneurs engaged in illicit activities, witnessed in the building's co-optation by drug dealers. Following the predictable symbiosis between drug dealing and other forms of vice, prostitutes patrolled the surrounding blocks.

Already by the 1980s, new patterns of development in older industrial neighborhoods were becoming apparent, with derelict spaces of the

industrial past reimagined around the residential and consumption requirements of a new class of urban residents, white-collar (or "no-collar")
workers in the growth sector of the postindustrial city. Moreover, many
were beginning to note the role that young artists played in this process.
Increased movement of artists into the Wicker Park area helped to set the
stage for the building's reanimation as a site of legal enterprise. The presence of the nascent artists' scene significantly influenced the location of
Urbus Orbis, and the opening of the venue helped push this scene into the
spotlight, laying the ground for its spectacular explosion in the 1990s.

This process draws upon the cumulative cultural identifications that
can attach to specific local places and to categories of place. The space of
the city is simultaneously real and imagined, organizing and organized
by both practical activity and cultural representations.[15] In the case of
Wicker Park, familiar elements of the modernist bohemia acquire new
meanings in relation to the structuring forces of the social field, with the
global dispersal of production and the heightened aestheticization of the
economy creating the context for former Chicago sweatshops being turned
into coffee shops, television studios, or the loft offices of Internet design
firms. As the immaterial attributes of commodities — their "sign value"
or cultural content — take on increasing importance in the pursuit of
profit,[16] so the new bohemia becomes more consequential in the city's
political economy. The concept of spatial practices thus alerts us to the
way that the neighborhood continues to organize social reproduction and
economic activity in a new era of urbanism.

The Industrial Neighborhood Takes Shape

When a motley prairie settlement on the shores of Lake Michigan incorporated as the city of Chicago in 1837, the area that would become
Wicker Park sat at its westernmost boundary.[17] This area was scantly
developed in the immediate decades that followed. The park that gives
the neighborhood its name was established in 1868; however, it was only
after the Great Fire of 1871 that the western fringe became the target
of intensive settlement, spared as it was by the legendary inferno.[18] The
postfire expansion of Chicago as a whole was truly stunning. "In 1870 the

city had a population of perhaps 300,000 living on only 35 square miles of territory," Abu-Lughod writes in *America's Global Cities*. "By 1893, the city's population would exceed 1.3 million living in an expanse of 185 square miles, making Chicago in that period the fastest-growing city in the history of the country, if not the world."[19] The fire cleared space more ruthlessly than could any planned project of urban redevelopment, setting the stage for the industrial metropolis to rise like a phoenix from its ashes.[20]

Chicago was the shock city of the frontier, expanding without the frictions associated with older periods of development, and this is one factor that has led both past and contemporary urban theorists to posit its ideal-typical status for a distinct sort of urban growth.[21] The expansion of the city was determined by existing levels of transportation and technology, combined with ascendant relations of production and exchange; and it was fed by specific processes of population redistribution, as European peasants flowed across the Atlantic.[22] Entering the twentieth century, Chicago's population further swelled with waves of African Americans migrating from the rural South to the metropolitan North.[23] In the 1920s, University of Chicago sociologist Ernest Burgess attempted to codify the two central tendencies of Chicago's morphology: the centripetal growth of the city as a progression of functionally differentiated concentric zones emanating from a densely built central business district, and, within these zones, the checkerboard of distinctive community areas into which competing racial and ethnic groups sorted.

Harvey Warren Zorbaugh wrote in the 1920s, "The Loop is the heart of Chicago, the knot in the steel arteries of elevated structure which pump in a ceaseless stream of three millions of population into and out of its central business district."[24] The limitations of transportation technology led to the clustering of industrial activity in dense pockets around the primary means of freight transport — water and railways. The elevated train's circular route encasing the central business district created the literal parameters of the Loop; as the twentieth century unfolded, the larger industries, particularly the hulking steel mills, dominated the South Side lakefront. As various immigrant groups clustered together in ethnic enclaves, their distributional patterns in the city also determined their sorting into occupational niches within the blue-collar economy. The means to transport the human freight of labor power to factory floors set limits and possibilities for community development and outward expansion.

If the Loop organized the city as a whole, Chicago was further subdivided into smaller, quasi-autonomous community areas, and the Chicago sociologists were pioneers in the study of these neighborhoods. While many theorists of the modern metropolis emphasized the alienating dimensions of urban life, Chicago sociologists like Robert Park and Ernest Burgess were attuned to how local communities organized both material and symbolic resources that eased the shock of displacement for new immigrant and migrant arrivals. Ethnic groups sorted into enclaves that provided local amenities, including churches, schools, newspapers, entertainment venues, and employment networks that were tailored to specific ethnic dispositions and stores of social capital. Park indicates of these communities, "Where individuals of the same race or of the same vocation live together in segregated groups, neighborhood sentiment tends to fuse together with racial antagonism and class interests. . . . The processes of segregation establish moral distances which make the city a mosaic of little worlds."[25] The differentiation of neighborhood areas was solidified by the material attributes of landscape, boulevards, rivers, or rail tracks. But it was the attachments of local residents that gave these communities their durability and symbolic power.

As Chicago expanded after the fire, Wicker Park was one such community area, even if Burgess overlooked it on his famous map.[26] The paving of Milwaukee Avenue in the 1870s was a key step in binding the Wicker Park district into the commercial concentrations of the loop, encouraging both wealthy and impoverished refuges from the fire to make permanent their settlements on the West Side. By the turn of the century, Milwaukee Avenue was "a major urban thoroughfare exploding with new businesses and carrying waves of ethnic immigrants out from the city's center."[27] Many well-to-do families, especially those of German-American businessmen, rebuilt mansions destroyed by the fire along the streets overlooking the park, and this housing stock survives in today's gentrified neighborhood.[28] At the backdoor of the park-front mansions, tenement housing also swelled, accommodating polyglot ethnic settlers. The legacy of this expansion continues to mark the local

Figure 2.2 Milwaukee Avenue in the Wicker Park Area Circa 1900. Photo from the Chicago Historical Society.

built environment, with the dense mixing of commercial development and widely variant housing stock.

Germans, Norwegians, Jews, Italians, and especially Poles crowded into the district over the course of the next several decades, with the population tide stemmed only in the 1930s, when the Depression finally slowed the Chicago juggernaut. The park-front gentry were embattled immediately by the crush of déclassé neighbors, and the postfire affluence increasingly gave way to working class grit. Dominic Pacyga and Ellen Skerrett write in their study of Chicago that by 1902, "the original mansions immediately surrounding Wicker Park had been turned into boarding houses and multiple family dwellings for the Polish working class."[29] The class relations of the industrializing city were contentious to be sure, and Wicker Park was, among other things, a seedbed of political radicalism. Among the earliest and most spectacular eruptions of labor unrest and violence, the Haymarket Square incident of 1886 took place to the near south of Wicker Park, and three of the five alleged "anarchists" who were executed in response hailed from the neighborhood. The park itself served as the terminus of their funeral procession.[30]

Heading into the twentieth century, Wicker Park experienced a fairly accelerated pattern of ethnic succession. However, the Polish workers who

crowded there would prove difficult to dislodge, dominating the neighborhood's ethnic identity throughout the first half of the twentieth century
and remaining a durable minority even in the century's latter decades.
This area around Ashland and Division came to be identified as Polish
Downtown or Polish Broadway.[31] According to Edward Kantowicz, this
area was 86.3 percent Polish by 1898.[32] This commitment of Poles and
other eastern Europeans to the district owes to cumulative local advantages, from ethnic churches to scattered sweatshops within walking distance
of workers' homes.

Wicker Park's local ecology organized spaces of work, leisure, and
private life for its Polish population. Ethnic associations, the Catholic
Church, and Polish taverns were all key neighborhood institutions that
gave shape to residents' daily lives and symbolic attachments.[33] Sociologists
at the University of Chicago produced exemplary works analyzing such
institutions throughout Chicago communities, including Helena Lopota's
midcentury study of voluntary associations in Wicker Park, which she calls
"Polonia."[34] Though Poles had long since been displaced as the dominant
residential grouping in the neighborhood, Wicker Park entered the 1990s
with a significant number of Polish restaurants and taverns still in existence, and the Polish Museum of America, founded in 1935, still operates
on Milwaukee Avenue.

The collective efforts of the Chicago sociologists to name and describe
the city's community areas are impressive and impressively durable in their
influence, so much so that Burgess' original map has taken on "official"
status in depictions of the city by politicians, researchers, journalists, and
local residents.[35] The projects carried out at the University of Chicago over
a period of several decades were united by a common object and launching pad, and by the shared theoretical perspective of "human ecology,"[36]
emphasizing the adaptation of the human organism to local environments.
This has become a canonical frame in urban scholarship, but one that has
also come in for much criticism. Of note is the oft-articulated complaint
that the focus on local systems obscures the importance of broader social
and historical forces. In other words, by examining Chicago's community
areas as "mosaic little worlds," Burgess and his heirs placed too much
emphasis on their internally coherent, bounded nature.[37]

The focus on self-contained neighborhood gives short shrift to the fundamental importance of developments in industrial production unfolding at a much larger scale. After all, it was the beacon of the smokestacks that guided white ethnic immigrants across the prairie. Yet in attempting to delineate the natural "laws" of urban process, the historically specific nature of the industrial economy is itself abstracted away. Writes Hise,

> Given the centrality of work and manufacturing in the city and in the lives of its inhabitants, it is all the more surprising that Park and Burgess were silent on the subject. Industrialization, the creation of large manufactories, [and] the transition from artisan to industrial worker — these were the transformative processes of the nineteenth and early twentieth centuries. Yet about these, they have little to say.[38]

It is just this sort of myopia that the frame of spatial practices seeks to avoid, analytically linking the near and the far, the immersion of local outcomes in broader social and economic forces. To avoid the trap of standard ecological approaches to the city, we must expand our frame to address large-scale trends in the life of the nation, trends that both structure and are structured by local outcomes.

Fordism

If America's foundational myth is that of the frontier and the cowboy in-dividualist, its twentieth-century economy increasingly demanded that workers live in cities and labor as if practically identical cogs in a machine. Taylor's scientific management and especially Henry Ford's assembly line were innovations in the division of labor extending the grip of rationality on the workplace.[39] The basic insight — that productivity could be dramatically increased by radically de-skilling labor and severing the design of products from their manufacture — was not exactly novel,[40] but the scale at which it would come to be implemented proved truly revolutionary.

Putting the rational principles of assembly-line production into practice posed two immediate problems. First, people must be persuaded (or compelled) to abdicate self-direction in the workplace and to labor as if automatons. Beyond that, someone had to buy the massive amount of practically identical goods that could now be produced, a historically unprecedented style of consumption. These were more than just problems of factory administration — the regulation of a social system of factory-style production necessitated corresponding *cultural* change. As Antonio Gramsci noted in the early 1930s, "In America rationalization has determined the need to elaborate a new type of man suited to a new type of work and productive process."[41] One might add the need for adoption of a new ethic of consumption, the "uniform mode of consumption of simplified products [that] is mass consumption."[42]

Maintaining workplace control was a serious problem for the early industrialists, one that they typically tried to solve via coercive measures such as firing diffident workers, withholding wages, and aggressively moving to counter the power of organized labor. Coercive measures were abetted by the surplus of impoverished new entrants to the city; drawing on his analysis of Manchester, England, as a proto-industrial metropolis, Engels argues that such cities organize intentional slums to house this reserve army of poor laborers, empty bellies providing for ready enlistees to serve as line-crossing scabs and strike-breaking thugs.[43] As the Chicago sociologists confirmed, the slum was a prominent feature of the early-twentieth-century industrial city.[44] Factory owners were further abetted by the ethnic and racial cleavages that undermined effective collaborations in the pursuit of better working conditions, cleavages that were inscribed on the "ordered segmentation" of industrial neighborhoods.[45]

Though many industrialists in the early twentieth century saw little advantage to rewarding their workforce beyond bare subsistence requirements, some attempted to balance coercive mechanisms with attempts at persuasion. Henry Ford is exemplary for the clarity of his program, recognizing that control in rationalized production did not end at the factory gates. Moreover, his novel generosity in instituting the $5 workday was more than an attempt to mollify the workforce. Ford saw that stunning advances in productive capacity meant little absent a pool of consumers, and it would in time become a matter of intense pride for autoworkers to own one of the cars that they had a hand in producing. Ford did not, however, trust his workforce to dispose of his largess responsibly, and he intruded on the nonwork lives of his employees in order to ensure that they consumed in as rational and disciplined a fashion as they produced.

Ford's twin innovations of the assembly line and the $5 workday are of sufficient importance that the term "Fordism" has been adopted by critical theorists to refer to the social regulation of mature industrial society.[46] Following Gramsci, Fordism signals a system of social organization in which rationally organized manufacturing is the dominant economic activity, involving a large portion of the workforce and effectively organizing a host of subordinate activities. Large cities were central sites of Fordist production as it unfolded through the first half of the twentieth century, hosting both factories and the office towers devoted to bureaucratized administration of the expanding corporate economy. Relatively high — and rising — wages would come to be viewed as a core component of the system's sustainability. In exchange for genuine participation in economic growth, workers were expected to evince docile submission to the requirements of rational production and to spend their incomes on the standardized goods that this system produced. In reality, such a total system proved exceptionally difficult to achieve, let alone maintain. Nonetheless, as an ideal-typical construction, Fordism is of central importance to our analysis.

The shock of the Great Depression appeared to derail the Fordist society before it ever got going. Indeed, it led many to question seriously the viability of the whole capitalist enterprise, as rational production outpaced viable solutions to the distribution problem. Like other industrial cities, Chicago was hit hard by the effects of the Depression, which exacerbated both labor unrest and ethnic conflicts in the battle over exceptionally scarce resources.[47] Extensive social programs were gradually implemented in response to mass unemployment and social suffering, and these would one day form core components of the mature Fordist society. But it took the Second World War to revivify the industrial economy. The United States emerged from this grand conflict with an unprecedented degree of geopolitical leverage and a monopoly position in the world economy that made its earlier industrial advantages pale in comparison.[48] Only under these circumstances would Fordism really come into its own, remaking American society and its material landscape.

A number of factors converged during this period. Large-scale corporate organization extended its breadth in the economy, but it was balanced with interventions by the state and organized labor in economic regulation. Postwar U.S. affluence brokered what Manuel Castells characterizes as

> a social pact between capital and labor which, in exchange for the stability of capitalist social relationships of production and the adaptation of the labor process to the requirements of productivity, recognized the rights of organized labor, assured steadily rising wages for the unionized labor force, and extended the realm of entitlements to social benefits, creating an ever-expanding welfare state.[49]

These arrangements provided for the creation of a relatively affluent blue-collar class, in which participants relied on standardized career trajectories for the reproduction of a style of life that included opportunities to participate in the mass consumer society.

The stability of postwar Fordism relied on a powerful cultural dimension — an ethic of conformity, the primacy of the nuclear family, and the embrace of rationalized consumption. In Chicago, Fordist affluence and immigration decline in the post–World War II period initially appeared to generate increased neighborhood stability. Working-class neighborhoods became bases of production in which a social pact between labor and capital was premised on uninterrupted economic expansion, undermining motives toward political action aimed at disrupting the system—that is, work stoppages. Ethnic conflicts likewise declined, other than the apparently intractable antagonism between blacks and whites.[50] The decline in European immigration reduced the pressures of invasion and succession, and postwar Chicago sociologists increasingly emphasized cooperation and "ordered segmentation" over the competition that concerned Park and Burgess.[51]

The massive steel mills on the south shore were the dominant manifestations of postwar Fordist accumulation in Chicago. William Kornblum examined Chicago's South Side in the 1960s, at what would in retrospect prove the tail end of local stability and comparative affluence.[52] In contrast to the early Chicago School neglect of political economy, Kornblum shows how neighborhood life was inscribed by the elements of Fordist production practices: routinized labor processes, vertical integration, standardized products, unionization, and a relatively affluent blue-collar workforce. "Space and time in South Chicago are organized in large part by the exigencies of steel production," he writes.[53] The abstract, mechanical time of the mill shifts, running production twenty-four hours a day, drove the organization of everyday neighborhood activities. Such spatial practices brought daily routine in line with the exigencies of the mode of production. The stability promised by these arrangements imbued neighborhood residents with what would prove a false sense of security in their relationship to space and economy. Even at the peak of Fordist prosperity, effects were far from uniform across Chicago's neighborhoods. Large corporations led postwar prosperity, and Wicker Park, where the enterprises were smaller and local, began to experience economic decline in advance of the crises that would beset Fordism after the 1960s. Moreover, the intervention of the state in remaking the material landscape of the city further undermined the viability of core city neighborhoods. The Kennedy Expressway cut through the northeast corner of the neighborhood, decimating the homes of some blue-collar residents and opening up a portal for suburban white flight. Fordism thus brought steady population erosion rather than prosperity to Wicker Park during the postwar decades. Nonetheless, manufacturing still was the primary source of employment in the neighborhood.

Limits to Fordism

The decades immediately following the Second World War would prove to be the high point of the Fordist model of social regulation. Indeed, this period has been called the "Golden Age of Capitalism," in which, Joseph Bensman and Arthur Vidich write, "capitalism had reconstituted itself and solved the internal and external problems brought about by the Great Depression and by fascism and Nazism."[54] This ostensibly pluralistic society led many thinkers to posit that the older class conflicts that marred the transition to industrial capitalism were a thing of the past.[55] For a brief moment, the basic tenets of the Fordist social pact seemed incontestable: the efficacy of the Keynesian welfare state, the rights of labor, and the ability of rationalized production strategies to ensure uninterrupted growth. But the triumphalism that would accompany postwar Fordism would ultimately prove short-lived — though not as brief, perhaps, as the 1990s triumphalism of the technology-driven "new economy" that I will examine in the chapters to come.

The economic devastation that the Second World War inflicted on the industrial infrastructure of Germany and Japan delayed the emergence of international competition but did not prevent it. The global hegemony secured for American industry in the postwar period rapidly dissipated as these economies were rebuilt; indeed, economic treaties such as the Bretton Woods accords failed to secure U.S. dominance under changing circumstances and increasingly contributed to ballooning trade deficits.[56] Fordist corporations were compelled to respond to declining profits by scrapping onerous arrangements such as the promise of secure employment and rising wages. Doing so depended in part on spatial reorganization, escaping the entrenched union resistance in older city neighborhoods and factory towns. The shocks engendered by capital reorganization strained the regulatory capacity of the welfare state, simultaneously increasing the number of petitioners for support and decreasing the resources available to meet those demands. These problems were particularly acute in center-city neighborhoods, where tax bases were decimated by white flight and deindustrialization.[57] By the 1970s, the fragile edifice of Fordism, reliant upon the balancing of corporate, state, and union power, was crumbling.

External and internal pressures increasingly exposed the rigidity of the Fordist system. David Harvey writes that "there were problems with rigidity of long term and large scale fixed capital investments in mass production systems that precluded much flexibility of design and presumed stable growth in invariant consumer markets."[58] Post-Fordist reorganization of capitalism can be read as the cumulative effect of efforts to circumvent these rigidities —Harvey characterizes these strategies as "flexible accumulation."[59] The older center city might be interpreted as itself a source of fixed-capital rigidity. If dense urban agglomeration was the spatial manifestation of industrial development through the first half of the twentieth century, in the second half this progressively ceased to be the case as new transport options, communication technologies, political strategies, and modes of organization increased the mobility of capital. Older cities now encountered a new economy, one that robbed them of many traditional advantages.

The Neighborhood in Crisis

Heading into the 1960s, Chicago still appeared to be an unstoppable industrial titan. Mayor Richard J. Daley's aggressive public works initiatives (the better to channel patronage and graft, cynics would say) led him to make the plausible claim that he poured more concrete than any mayor in American history.[60] It is ironic, then, that rationalized industry and activist government in many ways laid the foundations for older neighborhoods' decline. Ford's innovations in mass production and state interventions in infrastructure allowed for the automobile to open up the suburban frontier, transforming the nature of metropolitan growth.

Developments evident early on in Wicker Park would come to define the postwar fate of cities. With the era of mass European immigration at an end, intense racial polarization between black and white residents took center stage in Chicago, one impetus for suburban white flight.[61] What had been a polyglot of ethnic enclaves on the South Side increasingly gave way through the 1960s and 1970s to a massive, grossly segregated, and materially deprived African American community, its fortunes pummeled by the decline in the local mill economy. On the West Side, the older Polish population in Wicker Park was fast being overwhelmed by newer immigrant arrivals, from Latin America rather than the European countryside.[62] Fordist affluence was increasingly manifest in suburban homeownership, not the inner-city neighborhood.[63] Thus the golden age of postwar capitalism would prove to be the twilight of the older industrial neighborhood.

Once a necessary condition for large-scale industry, the density of the city was increasingly a disadvantage, forcing vertical factory construction that provided irritating assembly line friction.[64] Moreover, older blue-collar neighborhoods were sites of entrenched union participation, resistant to downward pressures on wages and more flexible work trajectories.[65] These factors could be circumvented only by breaking the perceived monopoly

Table 2.1
Manufacturing Employment in the Chicago Region, 1947, 1982, and 1992

Area	1947		1982		1992	
	Number	Percent	Number	Percent	Number	Percent
Chicago	668000	78	277000	37	187000	31
Suburban Cook County	121000	14	279000	37	235400	39
Collar Counties	64000	8	189000	25	185200	30
Total	853000	100	745000	100	607600	100

Source: Janet L. Abu-Lughod, *New York, Chicago, Los Angeles: America's Global Cities* (Minneapolis: University of Minnesota Press, 1999), p. 324

link between the core city and mass production. Table 2.1 shows the dramatic shift in manufacturing jobs from the center city to suburban locales in the Chicago region. Along with the reconfiguration of manufacturing production toward greater regional deconcentration, we also observe a substantial net decline, with a loss of 245,400 manufacturing jobs from the region, or 29 percent of the total. Improvements in transportation and the liberalization of trade laws now reconfigure the geography of production on a global scale. By the 1980s, multinational corporate interests were outsourcing labor-intensive sweatshop production in the apparel industry into Third World locales that serve as "export processing zones"[66] — the sort of sweatshop work that had once been done in neighborhoods such as Wicker Park. In the wake of this reorganization, Chicago bore the burden of white flight, a decimated public sector, rising poverty and crime, and intense racial polarization. Wicker Park and surrounding neighborhoods continued to attract immigrant populations after 1960, mostly Puerto Ricans and Mexicans; however, these new immigrants faced a fast-changing economic landscape.

Though 58 percent of the Hispanic workforce in Chicago was employed in manufacturing in 1970, that number had declined to 39 percent by 1991.[67] Wicker Park lost manufacturing jobs at a rate larger than that of the city as a whole; in the six-year period from 1977 and 1983 alone, 12,543 such jobs disappeared from the West Town area. By 1990, the West Town poverty rate stood at 32 percent.[68] This was not as severe as in some South Side neighborhoods, where the poverty rate exceeded 50 percent, but it was still substantially higher than the 21.6 percent rate for the city as a whole.

With the loss of industry came a loss of people. West Side's population steadily declined through the second half of the twentieth century; by 1980, the population of West Town was little more than 40 percent what it was in 1930.[69] Though much of Wicker Park's inherited built environment, including its park-front mansions, the Flat Iron Building, and the Northwest Tower, remained structurally sound, these were severely underused.[70] Young artists who moved into the neighborhood during the 1980s recall a dicey street environment characterized by drug dealing and prostitution. However, elsewhere in the city, particularly on the North Side lakefront, a new pattern of neighborhood development was emerging, as districts like Lincoln Park and Wrigleyville were being reconstituted to meet the needs of a new class of urban resident. Moreover, the Loop has proved to be a resilient site of intense economic activity in the postindustrial period, meeting new requirements for the global economy in such areas as finance, advertising, and other high-end producer services.

Chicago in the Global Economy

Extrapolating from Chicago's patterns of development in the industrial era, Chicago School sociologists assumed that the city would "naturally" concentrate its most important economic activities in the core, with growth subordinate to this "initiating and controlling"[71] metropolitan center. However, as Sassen notes, "The dispersal capacities emerging with globalization and telematics — the off-shoring of factories, the expansion of global networks of affiliates and subsidiaries, the move of back offices out of central cities and to the suburbs — led many observers to assert that cities would become obsolete in an economic context of globalization and telematics."[72] Demographers note the growth of newer Sun Belt cities following the deconcentrated housing patterns of Los Angeles — a dispersed urbanism apparently corresponding to the dispersed global economy. Mark Gottdeiner argues that the spatial practices of the contemporary economy now generate "deconcentrated urban realms"[73] dotted by nodes of commercial activity that Joel Garreau identifies as "edge cities."[74] In 2002, Michael Dear echoed Garreau's claim that "every city that is growing is growing in the manner of Los Angeles."[75] While marshaling forests of data, the explanation offered by LA School theorists like Dear and Soja mine theoretical perspectives originating in the humanities, reading postmodern fragmentation from the fragmented metropolis.

New models of urbanism that stress deconcentration and sprawl challenge the importance of old center cities; however, they tell only a partial story. Downtowns, including downtown Chicago, continue to generate activities that are hardly peripheral to the global economy. Countering deconcentration, the decades following the 1970s saw a resurgence of capital investment in the center city. Of the 1980s, Sassen observes, "This explosion in the number of firms locating in the downtowns of major cities during that decade goes against what should have been expected according to models emphasizing territorial dispersal; this is especially true given the high cost of locating in a major downtown area."[76] In *The Global City*, Sassen describes a dialectical process in the global economy in which consumer markets and certain types of production are geographically dispersed, while core administrative functions concentrate in dense urban areas.[77]

Her three case cities — New York, London, and Tokyo — are well known for having the largest concentrations of multinational headquarters in the world. But Sassen argues that the power of large corporations is "insufficient to explain the capability for global control."[78] The global city is a "postindustrial production site," with the object of production being innovations. The accelerated production of innovations in the city, from finance to technology to culture, often takes place in smaller enterprises that are linked to but not formally contained within the megacorporations that dominate the global economy. This is true of firms that provide specialized financial or legal services and of specialized service providers in fashion, media, advertising, and marketing. These highly specialized activities reap benefits from urban agglomeration and remain committed to large cities despite the higher associated costs.

Such postindustrial production is evident in Chicago, where the financial exchanges that once articulated midwestern agriculture and industry to a national economy now participate in aligning commodity and currency prices on a global scale. Still, the accomplishment of this shift relies on the specialized services that have accumulated along the LaSalle Street corridor over a century and a half of financial activity,[79] much as London's dominance in transnational banking is an inheritance of colonialism, even after the sun has long set on the British Empire. The comparative advantages of the global city are cumulative.

In fact, the role of cities as generative milieux for innovations of all sorts is crucial to understanding the reemergence of spaces that appeared to have outlived their usefulness in the wake of deindustrialization and the expansion of telematics and digital communication technologies, from the high-priced skyscrapers of the downtown to a gritty neo-bohemian neighborhood like Wicker Park. Grasping this requires attention to how cities organize contemporary production sites and markets, embedding complex economic activities that abet the advances of corporate globalization.

The growing importance of these services has a profound impact on the demographic makeup of cities like Chicago. In growth sectors of the global economy, education plays a key role, generating a new class of workers who influence urban policy and drive redevelopment in ways that diverge from the spatial practices of the Fordist, blue-collar middle class.[80] The weakness of the ecological functionalism that posits the city as integrated entity is exposed by new arrangements, with growth sectors of the city uniquely indifferent to the failing of the old industrial base. The contemporary city is increasingly a "dual city"[81] of extreme wealth in the growth sectors and postindustrial devastation elsewhere, with possibilities of assimilation from one sector to the other substantially curtailed. Still, the Wicker Park case suggests that even these spaces of devastation may in select cases still have a role to play in the future, reconfigured as a generative milieu for postindustrial innovation.

Cities such as Chicago are thus still production sites, despite the decline in manufacturing employment. It is a mistake to confine the concept of production only to durable manufacturing. Accounting for the ongoing economic vitality of Chicago requires new categories of production beyond those steeped in industrial capitalism. Emphasis on the production of culture extends global city theory. As Harvey notes, flexible production in the global economy emphasizes "quick changing fashions and the mobilization of all the artifices of need inducement."[82] These artifices employ old and new technologies to extend the penetration of capital into every available nook and cranny of social life, as total spending on marketing in its multiple forms increases alongside the degradation of real wages for material production.[83] Chicago historically has been home to a substantial advertising sector, including the massive firm Leo Burnett USA, and as Lilley and DeFranco demonstrate, this sector "spawn[s] numerous supporting arts-related jobs in geographically proximate arts-related enterprises."[84] It is precisely the clustering requirements of aesthetic production that anchor these enterprises in center cities. As we will see, the large advertising firms located downtown were integral to the development of the digital design sector in late-1990s Wicker Park, linking these "boutique" style firms into the globalized swirl of commodified signifiers.

Rethinking Neighborhood

Rather than consigning Wicker Park to obsolescence and decay, trends in global economic restructuring are a key part of its late-twentieth-century resurgence. The theories of the global city associated with Sassen and others can get us only partway to understanding this relationship. While effectively countering the LA School's overemphasis on sprawl as the signature outcome of globalization, the analytic lens of these approaches is directed at downtown office towers, telling us little about the fortunes of the sort of neighborhoods that the Chicago School so richly investigated during the industrial era. To the extent that the urban neighborhood is addressed, it is either as a cordoned-off warehouse of postindustrial despair or as a playground for materialistic young professionals. The status of the neighborhood as a postindustrial production site has seldom been seriously considered, at least until recently.[85]

William Julius Wilson offers a highly influential and nuanced perspective on the impact of basic economic changes on the fortunes of the inner-city minority poor,[86] providing an apt corrective to theoretical approaches that place the blame for inner-city poverty on the derelict moral hygiene of the poor.[87] Using Chicago's South Side as his base, Wilson indicates that postindustrial restructuring has isolated poor residents socially by cutting them off from both the culture and opportunities of the mainstream service sector. In charting these changes, Wilson thus shows how historical processes affect the production of space. Once the underclass neighborhood is achieved, however, this temporal dimension drops out. Space becomes reified, a postindustrial "natural area" apparently condemned in perpetuity to self-reproducing cycles of poverty. The inattention to *ongoing* dynamism is not trivial. Though Wilson's structural approach helps us to interpret developments in Wicker Park through the 1970s and 1980s that produced patterns of decay (admittedly less stark than those of the South Side), it cannot account for the neighborhood's 1990s turn.

Conversely, studies of gentrification show how some decaying postindustrial neighborhoods are revivified by new patterns of capital investment. Often these studies note the importance of artists in helping to initiate the process, though they typically do not make the connection to new patterns of post-Fordist enterprise emphasizing cultural production. Thus, a number of scholars have identified the usefulness of artists for property speculators on the Lower East Side,[88] but their emphasis on land-based entrepreneurs does not address the extent to which the practices of neo-bohemia contribute to other strategies of capital valorization in the postindustrial city.

Spatial practices linking the constitution of everyday life to labor relations and productive processes are subordinated in gentrification theories to the abstract space produced by relations of exchange. In effect, the artists are treated as placeholders; unwitting shock troops of gentrification paving the way for nebulously defined yuppies to follow on their heels. While surely correct as far as it goes, this mode of analysis does not adequately address changes in the urban occupational structure and the often-contradictory processes of the aesthetic economy. Instead, gentrification is also reified as a natural process, much like the Chicago School's old invasion succession model, only now in reverse. As in William Julius Wilson's underclass arguments, the "natural history" of gentrification ultimately results in an apparently homogenized and self-reproducing space devoid of contradiction.

In framing his work on the East Village, Neil Smith notes that "systematic gentrification since the 1960's and 1970's is simultaneously a response and a contributor to a series of wider global transformations."[89]

Nonetheless, he focuses almost exclusively on local property values and the actions of property speculators, urban political elites, and local boosters. His key concept is the rent gap: "the disparity between the potential ground rent level and the actual ground rent capitalized under the present land use" that makes "capital revaluation (gentrification) a rational market response."[90] Essentially, the theory argues that "capital" moves from the city in search of better speculative opportunities elsewhere. Abandoned city space then becomes increasingly derelict until it is cheap enough to suck capital back. At the very least, this model is too mechanical, failing to tell us why so many severely devalued spaces in the city fail to gentrify.

Though useful, such an approach does not illuminate the large-scale shifts in capitalist organization that lead center-city investments to take strikingly different forms in the postindustrial period. Because land rents are the key independent variable, other strategies of capital accumulation are not examined in their historical specificity. Without seriously analyzing the conditions for an aestheticized urban economy, we are left to derive "potential ground rents" by reading backward from gentrified neighborhoods, producing a tautological analysis. Smith does not seem to find it especially revealing that it was manufacturing and blue-collar jobs that left the city, while artists, nightclubs, information technology firms, and white-collar professionals are what now "flow in."

The simple dichotomy between capital and local residents, posited by Smith and many others, misses the multiple and contradictory ways that capitalist strategies intersect in and around a neighborhood. In fact, the homogenizing tendencies and escalating ground rents of gentrification conflict with spatial practices supporting aesthetic innovation and the conditions of reproduction for innovative labor. A relatively moderate cost of living is necessary to maintain the balance of cultural offerings that neo-bohemian neighborhoods provide, which include offbeat, experimental, and "alternative" fare; it is also necessary to ensure a reserve army of culture workers who live in the neighborhood area and are available to enter into flexible employment relations. Such contradictions imbue the neighborhood with an ongoing dynamism lacking in theories of either concentrated poverty or homogeneous gentrification.

As we have seen, the production of space is not only a matter of narrow property speculation; space is also essential to the organization and deployment of labor power and productive processes. This point seems easy to grasp when we look at the organization of the blue-collar neighborhood during Fordism, but it remains salient in the context of post-Fordist restructuring, where labor relations are complicated in new ways. These are not tertiary concerns. Ironically, former sweatshops are now being put to use in the manufacture of images for an aesthetic economy.

CHAPTER 3

Bohemia

I was a hopeful in rooms like this
when I was working cheap.
It's an old romance,
the Boho Dance,
it hasn't gone to sleep.
— Joni Mitchell, "The Boho Dance," 1975

"WICKER PARK IS NOT A PLACE — IT'S A STATE OF MIND!" So claimed the Internet-design entrepreneur David Skwarczek in 2001, as we shuffled up Milwaukee Avenue on an easily satisfied search for a bar in which to carry on a familiar conversation about art, ideas, and e-commerce. The comment came as we puzzled over whether his offices, on the corner of Milwaukee and Augusta, were in Wicker Park "proper," and if not, just where they were. In fact, Chicago's neighborhood boundaries are cultural conventions, lacking the formal status of a ward or even a census tract, and during the 1990s Wicker Park's boundaries seemed to expand southward, annexing territory from the less glamorous Ukrainian Village, at least in the real estate ads. Nonetheless, Skwarczek was suggesting something other than the subjective nature of territorial divisions in the city. As he elaborated over subsequent pints of beer, Wicker Park signals an ethos of creativity and experimentation, one which his website-design shop shares regardless of what side of Ashland Avenue it is on. Still, if not "a place," this ethos feature of modernity is the ongoing unsettling of "all social relations." "It pours us all into a maelstrom of perpetual disintegration and renewal, of struggle and contradiction, of ambiguity and anguish. To be modern is to be part of a universe in which, as Marx said, 'all that is solid melts into air.'"[3] Perpetual disintegration and renewal produces a cultural condition in which successive generations experience intensely their own historical originality, in contrast to the more or less mechanical reproduction of traditional social organization. On the other hand, as Berman points out, the general experience of flux and discontinuity has been with us long enough to produce recognizable patterns of response: "Although most . . . people probably experienced modernity as a radical threat to all their history and traditions, it has, in the course of five centuries, developed a rich history and a plenitude of traditions of its own."[4]

Bohemia is one such tradition. In fact, though bohemia has histori- cally been considered marginal within and in opposition to wider social currents (sometimes referred to, nebulously, as the mainstream), it is also an exemplar par excellence of central modernist impulses.[5] We can see the seeds of bohemian ideology in the Barbus artists of the eighteenth century,[6] and even more clearly in early nineteenth-century Romanticism.[7] These movements, largely carried by declining aristocrats, predict the antibour- geois and antimarket sentiments that recur throughout bohemia's history, despite shifts in class structure and market organization. They also con- tribute to the establishment of a distinctly modernist vision of the artist as lifestyle deviant. George Becker demonstrates how the social typology of the "mad genius" derives from the Romantics;[8] among such innovators, licentious sexual activity, a casual attitude toward conventional manners and hygiene, and a propensity to nurture extravagant moods with drugs and alcohol emerged as evidence of artistic eccentricity.[9] The notion that deviance is virtually coextensive with the artistic temperament has become among the most persistent characteristics of bohemia as a style of life.

Added in the early decades of the nineteenth century were the in- escapably urban connotations of bohemia. Urbanism as a way of life is imprinted upon the bohemian project, inflecting both practical strategies and aesthetic representations. Hence Baudelaire, that ur-*prince de bohème* and prototypical *flâneur,* diverges from Romantic precursors like Byron

This was particularly apt given that, like the scattered citizens of a defunct kingdom, many in this new class of penniless and passionate eccentrics had been violently dislodged from their own fixed place in the social cosmos.[18] Some were children of petty aristocrats on the wrong end of bourgeois revolution. Whatever their social origins, says Jerrold Seigel, all who wished to live from the arts were confronted by a game with new rules: "Like other social groups, artists and writers found themselves participating in a commercial market, a world of buying and selling from which they had been mostly excluded or protected before."[19] This new development was both liberating and terrifying, and the experience of insecurity has been not only endured but ideologically embraced in the traditions of *la vie de bohème*. "For the bohemian image has always been an intellectually uplifted version of the gypsy image as a community of chosen outcasts, claiming the spontaneous gift of creativity and willing to undergo great penalties to preserve their peculiar freedoms."[20]

Potential creators were pulled to the metropolis by the centripetal forces also drawing peasants from the countryside. Too numerous to be absorbed by the professions and ill suited by disposition to the sweatshops, those who would constitute the citizenry of bohemia settled into the low-rent districts of the city to ply their self-appointed trades. Writes Grana, "Paris not only attracted its literary *declasseés*, but manufactured them, and some contemporary accounts are realistic and off-hand enough to attribute the super-abundance of intellectual fervor to nothing more lofty than occupational frustration."[21] In this light, bohemia's fabled ennui emerges as a romanticization of the sting of failure, transformed into a brooding aesthetic in which one claims to reject what was already denied.[22]

Paris in the nineteenth century provides a splendid example of the contentious and contradictory nature of modernity. It was a place of arcades and barricades, of messy coalitions between the popular (working) classes and the bourgeoisie, and of a new breed of bohemians. As many have argued, the dynamics of the Parisian social field are inscribed upon exemplary works of art in this period of intensive modernist innovation: the mapping of class and spatial relations in the writing of Balzac and Flaubert,[23] Baudelaire's urban pastoral,[24] the spectacular street scenes of the Impressionist painters,[25] and the influence of the 1848 revolution on the art of Daumier, Delacroix, and others.[26] Paris was an object of artistic representation, available to be read with greater or lesser degrees of transparency; moreover, not only works of art but also the material relations and the ideological significance of artistic production were products of the "capital of modernity."[27]

In Paris, a variety of practical strategies and mythic associations of the modern artist coalesced, producing what has proved to be a durable model of *la vie bohème*, elements of which can be identified in a variety of subsequent urban locales, even as contexts of material production and social class relations have shifted. It will become clear that this model is thematic rather than dogmatic; despite the periodic appearance of this or that artistic manifesto, the reproduction of the bohemian life in new cities and periods derives from the enactment of transposable schemas, rather than rigid adherence to a system of rules.[28] Most generally, there is nurtured in bohemia the conviction that the artistic life constitutes a calling that encompasses the very soul of the producer (this despite the bohemian rejection of all utilitarian aspects of the Protestant ethic); that to produce art requires a commitment not only to the practical activity of creation but also to the artistic style of life. Notes Seigel of these proto-bohemians, "Ambitious, but without means and unrecognized, they turned life itself into a work of art."[29] In Paris, the artistic soul was externalized not only through paintings or poems, but also through social performance.

Though the bohemian prototype is by now (at least) as much a part of our recollection of nineteenth-century Paris as the barricades are, the position of bohemia in the European capital was undoubtedly tenuous. The birth of bohemia coincided with the economic and political revolutions that forged the great classes of nineteenth-century modernity, the bourgeoisie and the proletariat. The bohemians encountered both and identified with neither. Bohemian adherents maintained aristocratic commitment to distinction via cultural capital, mocking the tastes of the bourgeois and by and large ignoring those of the folk. This combined with a fetish for the hedonistic, the illicit, and the criminal gleaned from social and spatial proximity to the urban subproletariat and other "parasitical" elements. These were, after all, elements not unfamiliar to the Latin Quarter and Montmartre. As Grana notes:

In dominating the life of the provinces and becoming the main agency and clearinghouse for every sort of interest, from politics to business, luxury, entertainment and pleasure, Paris was bound to attract all those who had come to the city to make a living from its many incidental activities: the floater, the sharp, the playboy and the shadowy entrepreneur — people whose existence was essentially improvised, unconventional, ingeniously opportunistic, full of an easygoing lust for fun.[30]

Bohemia shared with these shadowy urbanites a culture of opportunism and a lust for experience, evidenced in licentious sexual norms and the liberal use of drugs and alcohol. The identifications with the urban underworld also imprinted itself on bohemian aesthetic representations; if young artists produced inevitably for a bourgeois market, there is evident throughout the canon of bohemian works a fascination with those who resisted location in the established class relations either of superseded feudalism or emerging capitalism.

Though bohemians made a virtue of marginality, many adherents and fringe followers nonetheless became great contributors to modernism in the arts and in political and social thought. Still, "bohemian" and "artist" are not simply substitutable terms; not all artists were bohemian, and for those who were, bohemia is often best conceived as a transitional moment — as Balzac indicated, "a stimulating interlude until the chance for real work arrives."[31] For most, that chance would never come, and, as we will see, the existence of a numbers game that inevitably produces far more losers than winners is among the most durable features of the artists' experience, continuing through the present day. What emerges in bohemia is a milieu that provides aspirants with both material and symbolic supports for the plying of such an uncertain trade, from inexpensive dwellings and tolerant cafés to a local status system with rewards that are unhinged, at least partially, from the vagaries of the market.[32]

Many in Paris were no doubt delusional with regard to their own talents, and for some the claim to artistic aspiration was no doubt a pretense. What was attractive was not just the practical activity of producing art, but also the license for unconventionality that *la vie de bohème* conferred. As

Ephraim Mizruchi suggests, "Because it is understood that some people can 'get away with more' than others, there is a tendency for pretenders to seek the cloak of protection associated with those committed to the tolerated lifestyle and thus derive sensual or material gains from this association."[33] Indeed, the modernist association between genius and madness documented (and deconstructed) by George Becker[34] made even the most outlandish behaviors not only tolerable but available to be read as markers of authenticity. Quite apart from the production of tangible works of art in painting, literature, or other media, we may view being bohemian as a mode of social performance, and the material space of bohemia as the stage upon which such performance gains its greatest efficacy as collective action.

This social performance constitutes a legacy as profound as that of the collected works of art produced by bohemians. David Brooks notes that "the French intellectuals set up ways of living that are by now familiar to us all."[35] This familiarity is not simply a consequence of widespread fluency in the works of Manet, Baudelaire, or Flaubert (although, as I would discover in Wicker Park, many contemporary artists are well versed in older traditions). Bohemia persists not only in works of art or as a set of free-floating ideas; it is a lifestyle with thematic elements that can be read through varied practical instantiations in a range of urban contexts since its Parisian origins. Both the idea of bohemia and its associated spatial practices have proved durable and portable, which is evident in cities throughout Europe and the United States. But as we will see, each bohemian eruption is both familiar and quite distinctive because of the material and spatial specificities that it encounters in a particular city at a particular time.

Across the Pond: New York and Chicago

In the United States, bohemia materialized most prominently at the turn of the twentieth century in the New York district of Greenwich Village, an avant-garde ghetto in a sea of cultural populism. Writes Christine Stansell, "When they thought of bohemia, turn-of-the-century Americans called upon an imagery of art, hedonism, and dissent from bourgeois life that originated in Paris in the 1830's."[36] As Stansell indicates, the Parisian prototype served as a model for action and self-understanding actively and consciously incorporated into the lifestyle of the New York avant-garde, and further provided a frame within which the loose aggregation in Greenwich Village has been and continues to be interpreted by others. Like Montmartre, the Village was an inexpensive district that attracted a variety of eccentric sorts. "Low rents (at least early on) allowed artists and writers to live there: indeed a come hither real estate market beckoned to anyone who fancied a life transformation."[37] As the twentieth century unfolded, the Village would be home to a cast of artists, political radicals, and lifestyle eccentrics whose collective efforts created a durable mythology, including Emma Goldman, Marcel Duchamp, Edna St. Vincent Millay, John Reed, Eugene O'Neill, Djuna Barnes, Mabel Dodge, Maxwell Bodenheim, and countless others.

If the Greenwich Village version of bohemia borrowed liberally from Paris' example, it nevertheless took shape against the distinctive backdrop of the exploding American metropolis, fueled by immigrants and expanding industry. As Caroline Ware notes, the urban bohemia is an embedded phenomenon, claiming autonomy but in fact sharing space with a host of other groups: "Italian immigrants and their children, Irish longshoreman, truck-drivers and politicians, Jewish shopkeepers, Spanish seaman, and a remnant of staid old American and German citizens made up the majority of the population and, in combination with their more famous neighbors, gave the life of this community its social texture."[38] While Parisian bohemians derided thrifty shopkeeper capitalism and engaged in their creative callings against the backdrop of arcades, boulevards, and occasional barricades, the New York avant-garde took shape in the context of the machine age: the birth of the skyscraper, the ascendance of industrial capitalism, and later the advent of new technologies of mass destruction in the world wars.

Nestled among recently arrived huddled masses whose yearnings to be free were frustrated by a reality of gross inequality, the Village was a hotbed of both artistic experimentation and radical politics, strains of which coexisted temperamentally with one another. The radical left-wing journalist John Reed made a base there, as did protofeminists like Margaret Sanger and Emma Goldman. The dashing Max Eastman published *The Masses* in the Village, a socialist and fiercely antiwar publication that nonetheless in fine bohemian fashion chafed routinely at the imposition any sort of party dogma. In fact, *The Masses* would emerge as at least as interested in literary experimentation as genuine revolution, drawing on the range of talents concentrated in the district.[39]

Generally speaking, though, bohemianism made a poor match with conventional left-wing class politics. While modernist artists fall short of postmodern successors in making a fetish of capitalist commodities, they remained nonetheless mesmerized by the spectacles of capitalism, as well as by the recklessness of the urban underworld. As Stansell notes of the American bohemians mingling among immigrant laborers, "these people were not slummers and they were certainly not philanthropists; they came to revel and discover, not aid and uplift. . . . [They prided] themselves on a hedonistic familiarity with the city and its gaudy and besmudged riches."[40] Compared to these diversions, the everyday struggles of the shop floor were altogether too tedious and, with few exceptions, were ignored.

If Greenwich Village was the most storied of American bohemias in the early twentieth century, it was not the only spot in a major U.S. city to follow the pattern of an artistic coterie nestling among the working classes. In *The Gold Coast and the Slum,* Harvey Warren Zorbaugh maps the development of Chicago's near North Side, exposing the striking juxtaposition of the privileged Gold Coast to the impoverished areas just west of State Street: "a nondescript world of furnished rooms; Clark Street, the Rialto of the half-world; 'Little Sicily,' the slum."[41] It is in this remarkably heterogeneous district that Zorbaugh gleans Chicago's own version of bohemia, called Towertown after its proximity to the landmark Chicago Water Tower: "Ironically enough, the last remaining landmark of the sternly moral, overgrown village that was Chicago before the fire becomes the symbol of the bizarre and eccentric divergences of behavior which are the color of bohemia."[42]

Zorbaugh saw the Towertown bohemia as already on its way out when he wrote, "The passing of Towertown, as of the bohemias of Paris and New York, is incident to the march of the city."[43] Such an assessment seems virtually obligatory, so often does it recur. Still, time has been particularly cruel to this oasis of prairie bohemianism, which is today almost entirely forgotten. The Water Tower remains a prominent landmark, surrounded by the malls and boutiques of Chicago's busiest shopping district. And Chicago, for all its exemplarity as a twentieth-century industrial titan, remains at best a poor stepchild to New York in terms of its cultural reputation.

No doubt Chicago, as Stansell suggests, "had a strong claim to the spirit of the age: with its huge polyglot population, mammoth industrial base, and gorgeous skyscrapers, it was the newest city in the age of the new, the shock city of the early twentieth century."[44] Still, so thorough is New York's perceived dominance of the more esoteric pursuits of the mind, it's easy to forget that in the early twentieth century, Chicago was not only a center of industry, but also challenged New York in the literary arts. H. L. Mencken claimed that Chicago was the emerging literary capital of the United States, writing in 1917 that in all likelihood the writer who "is indubitably American and who has something new and interesting to say, and who says it with an air . . . has got some sort of connection to Chicago."

Mencken argued that writers like Theodore Dreiser and Frank Norris "reek of Chicago in every line they write." The literary naturalism practiced by the Chicago writers recalls classic themes of nineteenth-century French literature; Dreiser's Sister Carrie, a child of the provinces mesmerized by the gaudy splendors of Chicago's department stores, is reminiscent of the novitiates depicted by Balzac and Flaubert. Like past bohemian authors, the Chicago naturalists gleefully incorporated various affronts to bourgeois propriety, including adultery, gambling, and drug use, but they also paid somewhat unusual attention to the actual working lives of the urban proletariat. Upton Sinclair, Edna Ferber, and Carl Sandburg all used the near North Side, nestled in the shadow of the skyscraper, as a perch from which to observe the stark contradictions of the industrial city, unfolding within a few blocks: the privilege of the Gold Coast, the dissolution of the slum, the tawdry diversions of Clark Street's gambling salons. Few today remember Towertown as the stomping grounds of Chicago's "literary left bank,"[45] but the efforts of those authors continue to influence Chicago's distinctive cultural style.

To focus on these famous names largely misses the point, though, even if it is an unavoidable temptation. In Paris, in New York, and in Chicago, the luminaries, Baudelaire, Manet, Duchamp, Sinclair, Sandburg and so on, were always soundly outnumbered by those whose destiny was obscurity. Impossible to forget, since we never knew of them anyway, they provide nonetheless the essential service of creating the critical mass that makes bohemia viable (even if always nestled in the larger mess of the "everyday" city). And as Zorbaugh points out, what distinguished Towertown most of all is the style of life that it licensed, one by now familiar to us: "Towertown . . . is largely made up of individuals who have sought in its unconventiality [sic] and anonymity — sometimes under the guise of art, sometimes not — escape from the conventions and repressions of the small town or the outlying and more stable communities of the city."[46]

As with Paris, adherents evinced a strong streak of fidelity to the autonomy of the aesthetic; nonetheless, they could not entirely escape the practical realities of the cities in which they were embedded. From the publication of *The Masses* in Greenwich Village to the muckraking impulses of Towertown's literati, the conflicts and contradictions of industrial capitalism also marked the contours of bohemia. Still, any critique of the city was mixed with thrill at the liberation offered by urbanism as a way of life; if the Chicago sociologist Louis Wirth saw the heterogeneous metropolis as alienating,[47] bohemians in Chicago and New York were adept at creating their own subcultural alternative within it.

Bohemia v. Mass Culture

The end of the Second World War brought further changes to the urban landscape, to U.S. society more generally, and to the next generation of bohemians. In the last chapter, we saw the complex of changes that accompanied the postwar period: the end of mass European immigration, the fruits of large-scale interventions by the state into public works projects, the incipient pull of the suburbs enabled by both state intervention and technological innovation, the growing state regulation of capital, the increased institutional legitimacy of organized labor, and the apparent pacification of past industrial conflicts by new norms of distribution incorporating the laboring classes into a system of mass production *and* mass consumption. On the heels of the Great War and the Great

Depression, a new American society was emerging, one rooted in the conflicts of the past but willfully determined to put all that behind it.

Leftist critics during this period of postwar Pax Americana were less likely to emphasize the material deprivation of the working class, instead focusing on the soul-deforming effects of mass culture. The impersonality of bureaucracy and the assembly line, dominant forms of the hyperrationalized organization of Fordism, alarmed even centrists like David Riesman and William H. Whyte, who felt that the frontier spirit of American individualism had been replaced by a committed ethic of conformity. This sentiment resonated enough at the time to make books like *The Organization Man* and *The Lonely Crowd* best-sellers (although perhaps some buyers were just following the herd). The exiled neo-Marxist critics of the Frankfurt School extended such observations into a total critique of American mass society.

T. W. Adorno and Max Horkheimer argue that culture had become just another Fordist commodity, following the same principles of production: "The technology of the culture industry [is] no more than the achievement of standardization and mass production, sacrificing whatever involved a distinction between the logic of the work and the logic of the system."[48] Often overlooked in Adorno and Horkheimer's analysis of the culture industries is a critique of the Fordist production of space and its attendant standardization of the built environment in downtown concentrations of capital as well as the nascent suburbs. Write Adorno and Horkheimer:

> The decorative industrial management buildings and exhibition centers in authoritarian countries are much the same as anywhere else. The huge gleaming towers that shoot up everywhere are outward signs of the ingenious planning of international concerns, toward which the unleashed entrepreneurial system was already hastening. . . . [T]he new bungalows on the outskirts are at one with the flimsy structures of world fairs in their praise of technological progress and their built in demand to be discarded after a short while like empty food cans.[49]

Thus, the stultifying routinization of the workplace seeps into the other spaces of lived experience, both the material space of the city and the expressive space of culture. Not only the Frankfurt theorists read the rationalization of city space as an assault on culture and creativity. In her hugely influential book *The Death and Life of Great American Cities*, Jane Jacobs likewise decried modernist city planning for destroying the vibrancy of the older urban neighborhood. For Jacobs, this life emerged organically from the heterogeneity of the city sidewalks, and she found her model for the living city at her own doorstep in Greenwich Village, the most storied of American bohemias.

We can see how such criticisms resonate with the ideology that has defined bohemia since its Parisian origins. The rationalized organization of labor and commerce were anathema to bohemian sensibilities, and bohemian districts emerged as privileged sites of resistance to this perceived sterility. The modern metropolis may have been a product of the bourgeois revolution, but for the bohemian fringe it was also a condition of possibility for the articulation of an alternative style of life. Heterogeneity, ephemerality, jarring juxtaposition, disorder — these are the qualities that have long organized the fantasies of the artist in the city. What is intriguing is the convergence of these bohemian ideals with the cultural critique of mass culture lodged by neo-Marxist intellectuals in the 1950s.

The urban hipsters of the beat generation provide the best lens into the articulation of postwar bohemianism. While best known through the works of a small cadre of writers, the beat moniker can in fact be applied to fairly durable urban subcultures in districts such as New York's East Village and Greenwich Village,[50] Los Angeles' Venice Beach,[51] and San Francisco's North Beach,[52] maintaining coherence as "the only significant new group of rebels in America"[53] for nearly two decades before being displaced by the 1960s hippies. The beats registered their discontent with U.S. culture in the Fordist postwar period of mass production, when "middle class" aspirations seemed to follow the standardized model of suburban home ownership and gray flannel suit conformity that is the antithesis of a bohemian urban aesthetic.

Given that the best known work from the beat writers is Kerouac's *On the Road*,[54] one might be tempted to view their movement as liberated from the urbanism that I have argued is integral to the definition of bohemia. This would be a facile interpretation; in fact, the growth of the suburbs seems if anything to have sharpened the place of the city street in the imaginary of authentic subcultural resistance. While mobile, the beats nevertheless used the road mainly to connect two coastal nodes of urban bohemia, in New York and San Francisco (with memorable detours in *On the Road* to New Orleans and Denver).

Not surprisingly, the cultural avant-garde sought solace in those spaces of social life that still confounded the Fordist utopia of standardized consumption. Previous generations of American bohemians shared space with white ethnics; the beats sought proximity to African Americans and were entranced by black cultural innovations, especially jazz (as well as by the antimaterialist strains of Eastern mysticism). They did not so much view those groups that the Fordist social pact still excluded —nonwhite populations, homosexuals, drug addicts, and outlaws — as victims of material oppression. Instead, writers like Kerouac, William Burroughs and Allen Ginsberg lionized them as the last bearers of authenticity in the mass consumer society of tasteless automatons. Compared to the sanitized spaces of Fordist privilege, those derelict enclaves where such subaltern identities were most often confined were objects of bohemian fantasy, much as the squalid street life of Montmartre had been. Thus Kerouac could without irony express his envy of the "ecstatic" Negroes inhabiting Denver shantytowns.[55]

Norman Mailer, who like Kerouac and Ginsberg was Ivy League–educated but eager to identify with the presumed netherworld of the city, penned a manifesto of beat-style hipness, "The White Negro," that is an unparalleled document for examining both the hypermasculinity and racial fetishization of bohemian yearnings. The hipster, following Mailer's famous formulation, repudiates "mainstream" society by aligning with the city's outlaw fringe. Of becoming hip, Mailer wrote:

> The only life giving answer is to accept the terms of death, to live with death as immediate danger, to divorce oneself from society, to exist without roots, to set out on that uncharted journey with the rebellious imperatives of the self. In short, whether the life is

criminal or not, the decision is to encourage the psychopath in oneself, to explore that domain of experience where security is boredom and therefore sickness, and one exists in the present . . . the life where a man must go until he is beat, where he must gamble with his energies through all those small or large crises of courage and unforeseen situations which beset his day, where he must be with it or doomed not to swing.[56]

Mailer's image of the "Square" is emphatically of a piece with the organization man, "trapped in the totalitarian tissues of American society, doomed willy-nilly to conform if one is to succeed."[57] But while his account is embedded in the conditions of postwar Fordism, its resonance extends back to the bohemian complaint with bourgeois sterility dating to Baudelaire's Paris. Moreover, we will see in the next chapter that the misogyny, racial fetishization, and bravado expressed in this essay continue to be recognizable in contemporary Wicker Park, which produces its own versions of the "existential hipster."

The beats did not have a corner on such fascination with marginalized groups and their urban niches. Novelist Nelson Algren lived in Wicker Park during the postwar years, setting such works as *The Man With the Golden Arm* there. Within this working-class Polish neighborhood, Algren typically deemed only the grifters, prostitutes, gamblers, and drug addicts as worthy of much attention. That he considered their authenticity most credible is demonstrated by his backstreet tours with Simone de Beauvoir, in which he promised to show her the "real" Chicago.[58] And as we will see in the next chapter, Algren, while decidedly second tier in the American literary tradition, also proved to be quite influential for the 1990s bohemians of Wicker Park.

Moreover, even Marxist critics, disenchanted by the failure of the laboring classes to evince revolutionary agency, began to turn wistful yearnings in the direction of the subproletariat. Herbert Marcuse, having diagnosed postwar society as a total mechanism eclipsing critical thought, tentatively identified the excluded as a last hope for resistance: "Underneath the conservative popular base is the substratum of outcasts and outsiders, the exploited and the persecuted of other races and other colors, the unemployed and the unemployable. . . . [T]heir opposition is revolutionary even if their consciousness is not."[59] This interpretation shares with bohemian inclinations a patronizing relationship to the subaltern; if the dispossessed are indeed the last bearers of authenticity in a world of depraved conformity, it still remains for the intellectual avant-garde to interpret and translate the critical potential in their brutalized psyches.

The 1960s: Mass Counterculture

The 1950s are popularly recalled as a period of staid and comfortable conformity within what is otherwise a tumultuous twentieth century, regarded as either idyllic or horribly stultifying depending on one's taste. The 1960s are an entirely different matter, a period when new social movements apparently launched a full frontal assault on the pieties of the comfortable middle class. The rise of identity politics in the 1960s seemed to mark the return of conflict to the political arena. These new social movements, which derived much inspiration from both the beats and Marcuse, were not a return to the old left/right antinomies that Daniel Bell and others suggested had been exhausted.[60] While it was clear that the Fordist social pact excluded women and black Americans among others, and produced a homogenized culture that the rising countercultural youth movement objected to, social class and the critique of capitalism no longer occupied the core of left-wing criticism.

The New Left politics of the 1960s converged with the rise of a countercultural identity, perceived by adherents as profoundly generational and historically original even while borrowing liberally from bohemian traditions (including conscripting beats like Ginsberg and Neal Cassidy into the movement). While the New Left and the counterculture are often considered as distinct from one another, David McBride argues that during the 1960s, "the practice of political and aesthetic radicalism in everyday life meshed considerably."[61] This meshing took place in distinctive sites of subcultural activity; rural communes may have proliferated, but urban districts retained crucial importance. During the 1960s, the avatars of the counterculture staked a claim on their own city spaces as sites of new bohemian fantasy, notably the Haight-Ashbury district in San Francisco and Berkeley's Telegraph Avenue. In these storied districts, both acid tests and protests were the order of the day.[62]

Still, if the bohemian impulse had been contained in a few eccentric urban districts since Parisian origins, theorists as disparate as Allan Bloom, Fredric Jameson, and Daniel Bell assert that the 1960s constituted a crucial breaking point, in which bohemianism progressively became a defining force in the culture generally.[63] As Bell puts it, "The adversary culture has come to dominate the cultural order."[64] The main carrier of this impulse, most agree, is the popular youth culture that exploded in the 1960s, especially rock 'n' roll. Todd Gitlin points to the chart-topping popularity of Barry McGuire's "Eve of Destruction" in 1965 as an indicator of a decidedly changed mood, in which youth everywhere embraced the disaffected stance that once was the property of a handful of urban malcontents.[65]

The far-ranging political and cultural implications of that eventful decade were accompanied by more subtle developments in the economic realm. Fordist affluence enabled both expanded market power and increased levels of education for the children of the middle class, key features in the progressive shift from the suburban homemaker to the fashion-forward youth as the model consumer. At the same time, the foundations of Fordism were cracking, even if those cracks eluded the inspection of most of the period's experts. The limits of mass standardized production for sustaining domestic growth would culminate in the economic crises that wracked the 1970s. New strategies emerged in the face of this crisis, contributing to a system of flexible accumulation that includes a new geography of material production.[66] Flexible accumulation also generates an expanded role for culture in the world of commodity production. This extension of capitalism's symbolic economy has, in fact, come to be a signature feature of the postmodern epoch, as Perry Anderson observes:

> Culture has necessarily expanded to the point of where it has become virtually coextensive with the economy itself, not merely as the symptomatic basis of some of the largest industries in the world — tourism now exceeding all other branches of global employment — but much more deeply, as every material object and immaterial service becomes inseparably tractable sign and vendible commodity.[67]

Radical though they appeared, the stylistic innovations of the '60s counterculture have played a part in the massive extension of consumer culture since. Though the bohemian fringe has always disproportionately affected modernist cultural innovations, after the 1960s the importance of the cultural marketplace to the economy more generally would surely elevate the economic relevance of the older bohemias' postmodern heirs.

Postmodern Bohemia

If the 1960s were indeed the tipping point, the growth of the cultural marketplace has been spectacular since. While many note increased production in television, film, literature, and popular music, equally relevant is the increase in the class of cultural producers that this volume implies. Always a trenchant observer of social trends, Bell saw in the 1970s a phenomenon that has only become more pronounced: "There has been an evident change in scale. Even though tiny by comparison with the numbers of the total society, the present cultural class is numerous enough for those individuals no longer to be outcast, or a bohemian enclave, in the society."[68] Counting artists is an inexact science, and the number of artists alone does not capture the breadth of economic activity organized around symbolic production, which includes the extensive technical expertise of new media production, marketing, and design. Still, an examination of data on artists, writers, and performers in the twentieth century is revealing. Through most of the century, proportional rate of growth in this population was steady but unspectacular — until 1970. Thus, from 1900 to 1970, the number of artists, writers, and performers per 100,000 of the population in the United States went from 267 to 385, an increase of 44 percent. By 1999, the proportion had jumped to 900 per 100,000, an increase of 237 percent in just three decades. In absolute terms, the number of artists, writers, and performers grew from 791,000 in 1970 to almost two and a half million in 1999 (table 3.1).

This startling increase lends credibility to the idea of an expanded cultural economy that impacts employment as well as consumption. What are the implications for the contemporary relevance of bohemia? Surely, the population of cultural aspirants (and groupies) who have

Table 3.1

Artists, Writers, and Performers in the United States, 1900–1999

Year	Artist's Population	US Population	Per/100 Thousand
1900	203,000	76,094,000	267
1910	288,000	92,407,000	312
1920	291,000	106,461,000	273
1930	406,000	123,077,000	330
1940	439,000	132,122,000	332
1950	524,000	152,271,000	344
1960	608,000	180,671,000	336
1970	791,000	205,052,000	385
1980	1,284,000	227,224,000	565
1991	1,957,000	249,464,000	784
1999	2,454,000	272,691,000	900

Sources: 1900–1960 from US Bureau of the Census, Historical Statistics of the United States, Colonial Times to 1970 (1976). 1970–1991 from US Bureau of Labor Statistics, Employment and Earnings as Reported in the US Bureau of the Census, Statistical Abstract of the United States. 1999 from the US Bureau of Labor Statistics, Occupation and Employment Statistics (available online). Thanks to Richard Florida and Kevin Stolarick.

traditionally passed through bohemia now exceeds the carrying capacity of the relatively small number of enclaves that hosted them in the past. But to say that they can no longer be contained in Greenwich Village is not to suggest that artists are somehow randomly distributed geographically. They remain disproportionately committed to center-city living, even as center cities struggle to weather the effects of economic restructuring. In fact, Ann Markusen shows that artists are an especially fast-growing population in cities not traditionally associated with cultural production, including Minneapolis, Cleveland, and Detroit.[69] Bohemia, if defined solely as an urban artists' district, has only become more frequent in the postmodern period.

But does this ubiquity itself spell bohemia's demise? After all, we have explored how bohemia signals not just a container for artists and their friends, but also denotes a distinctive style of life conditioned by both spatial and social location. With the advent of the expanded cultural economy, it is possible that perhaps something new has taken shape, a hybrid incorporating elements of bohemia and the older mainstream, as implied by David Brooks' catchy neologism, BoBos (bourgeois bohemians),[70] or more dramatically still a "big morph" that, as Richard Florida argues, produces a "new mainstream."[71]

In fact, Daniel Bell was arguing by 1976 that the mainstream culture had been fatally contaminated by the hedonism of bohemian life.[72] According to Bell, the mass counterculture, with its emphasis on self-actualization and immediate gratification, constituted a "cultural contradiction" in capitalism, undermining respect for authority and commitment to the Protestant work ethic. A quarter century of hindsight suggests that the post-1960s consumer culture is not the dire threat to capitalism that Bell imagined. Indeed, insofar as it heightens the importance of fashion and cultural consumption, the inheritance of the 1960s counterculture opened up new avenues for cultural industries to pursue profit.[73] In other words, rather than planting the seeds of capitalism's own destruction, the 1960s mass counterculture anticipated key elements of a solution to Fordist rigidity.[74] For Brooks, BoBos resolve the contradictions that Bell identifies, working like the bourgeoisie and consuming like bohemians, and the (capitalist) beat goes on.

While Brooks and Bell focus on the realm of consumption, Richard Florida turns his lens on the occupational structure of the "creative economy," in which the important roles are filled, naturally, by members of the creative class.[75] Florida defines this class broadly, and he includes in it such well-established and seemingly staid professions as finance, medicine, and the law, fields that would seem to have very little to do with bohemian dispositions. But the people who really interest Florida are the programmers, designers, and entrepreneurs who were at the center of the extravagant, though short-lived, Internet boom during the late 1990s. With their long hair, tattoos, and penchants for living in artists' neighborhoods and moonlighting as rock musicians (Florida describes his epiphany upon learning that the musicians entertaining at an Austin, Texas, technology conference were also "high tech CEOs and venture capitalists"),[76] these sorts do indeed suggest something of the bohemian. But to Florida's own surprise, in interviews with members of the creative class, "they bridled at the suggestion that they were in any way bohemian."[77] Taking these informants at their word, Florida goes on to offer the interpretation that "bohemians were alienated people, living in the culture but not of it, and these people didn't see themselves that way."[78] On the basis of this evidence, Florida concludes that bohemia has morphed into something else, a new mainstream in which creativity, nonconformity, and lust for visceral experience can coexist comfortably with grueling work schedules and the pursuit of massive profits.

Florida's relentlessly cheerful account of a creative economy in which alienation is a thing of the past,[79] at least for the talented, reads like a relic of the last years of the Clinton era and "new economy" triumphalism.[80] Though the brave new world of the digital economy produced a few high-profile superwinners, it also produced a number of losers even before the market crashed. High-tech enterprise, essentially technologically souped-up graphic-design firms, did penetrate Wicker Park, conscripting some number of local artists into their ranks of employees and subcontractors. These Wicker Park residents were not immune to the fantasy that they could be creative, edgy, and rich all at once, but the reality was far more pedestrian, with long hours, mediocre wages, and extraordinary vulnerability.

Buttressing his claim that alienation and self-destruction are passé, Florida indicates that, among performing artists at least, self-destruction was replaced in the 1990s with physical fitness: "A number of middle aged rock stars, such as Bruce Springsteen and Madonna, now appear much fitter than when they started out. Some musicians have bigger biceps than pro athletes did forty years ago; if Bob Dylan were to come along today, his agent would probably send him to the weight room."[81]

This generalization is suspect. Self-destructive impulses did not drop out of the artistic culture in the 1990s. Leaving aside the fact that Dylan released several critically acclaimed albums during the decade, Florida ignores the many musical artists for whom self-negating activity such as drug abuse remained central features of their personae, such as Kurt Cobain, Courtney Love, and Layne Staley, just as he ignores the generally nihilistic tone of the decade's ascendant genre, rap. At least some contemporary urban hipsters have inherited from Mailer's prototype darker desires than the yen for bicycle paths and street musicians that Florida attributes to creative-class experience hounds. Surely health clubs proliferated in the 1990s, but it was also the decade of heroin chic.

Bell, Brooks, and Florida make grand gestures, and their reach exceeds their grasp. Moving too far from actual neighborhoods in their examination of postmodern bohemia, they lose hold of the concept altogether. But their insights still help us to understand what is historically distinct about new bohemian spaces such as Wicker Park. Bell's analysis of the lust for entertainment in postindustrial society illuminates the dramatically expanded market for cultural commodities. This expanded market elevates the place of the new bohemia in the networked geography of cultural production. Though Brooks' BoBos are not entirely unprecedented — past elites were known to slum in Montmartre, Greenwich Village, Harlem, and so on — the willingness of educated professionals to eagerly take consumption cues from urban artists does appear more widespread, in part because such professionals have become a much larger share of metropolitan populations. As we will see, this attraction helps to explain why a neighborhood like Wicker Park becomes not only a target of gentrification, but also a bohemian-themed entertainment district where patrons are not starving artists but rather affluent professionals. Moreover, the Wicker Park case provides support for Florida's argument that new labor contexts are no longer anathema to either the dispositions or the competence of bohemian artists, though this does not, as it happens, spell the end of alienated labor.

New bohemian districts are increasingly common features of the contemporary urban landscape, their importance only enhanced by post-1960s mass counterculture. They are recognizable via the cumulative identifications generated in and around older bohemias. These neo-bohemias make unprecedented cultural and economic contributions to the broader social system without ever losing their distinctiveness within it. We can see the replay of key bohemian themes in Manhattan's downtown scene of the 1980s, nurturing new bohemian icons like Basquiat and pop stars like Madonna. Beyond New York City, new bohemias have flared up in recent decades in Seattle, Dallas, Detroit, Atlanta, Omaha, and, of course, Chicago. Anchoring ourselves in the Wicker Park case, and treating bohemia as a mode of spatial practices combining place and mind-set, rather than as a category of individual or as *only* a cultural style, we can make empirically informed judgments about the nature of bohemia's ongoing importance. Significant continuity exists in the contemporary artists' neighborhood; nonetheless, the new bohemia unfolds within a dynamic urban landscape, and shifts in the social field serve to produce something both familiar and, in important ways, quite new.

PART II

Postindustrial Bohemia

CHAPTER 4

Grit as Glamour

In murky corners of old cities where
everything — horror too — is magical,
I study, servile to my moods, the odd
and charming refuse of humanity.
> — Charles Baudelaire, *Les Fleurs du Mal*, 1857

"Welcome to Wicker Park"

WICKER PARK'S FLAT IRON BUILDING occupies the southeast
corner at the intersection of North, Damen, and Milwaukee,
in the shadow of the Northwest Tower and across the street
from the Urbus Orbis/*Real World* building. Its cast iron
façade evokes New York's SoHo district; the diagonal path of
Milwaukee Avenue cuts the structure's vee shape. Like so many
buildings in the neighborhood, the Flat Iron was originally
a home to light industry, now long gone. The ground-level
units house boutiques, art supply stores, restaurants, coffee
shops, and performance venues. Since the late 1980s, the
upper floors' studios have mainly rented to local artists — at
submarket rates, insists Bob Berger, the building's majority
owner. As the block was shifting from a blighted stretch of
postindustrial decay to a hipster haven in the early 1990s, one
of these artists placed a poster into a second-floor window,
a still from the now-classic 1976 Martin Scorsese film *Taxi
Driver*. It shows Robert DeNiro as Travis Bickle, the titular

Figure 4.1 The Flat Iron Building. Courtesy of Bob Berger.

taxi driver. Sporting a Mohawk, Bickle grins maniacally as he points two guns out onto the street. This poster remained in place for years, silently presiding over the transitions of the 1990s. Over the still, a hand-lettered sign was inserted, reading "Welcome to Wicker Park."

The poster is a distinctive marker of meanings attached to the neighborhood, an advertisement for Wicker Park as neo-bohemia. The redevelopment of Wicker Park intercepts the trends of postindustrial disinvestment and decay that motivated dystopic representations of the city such as those of *Taxi Driver*, even as many new residents remain enamored with the aesthetics of the gritty urban street. New economic activity hardly signals return to past foundations of relative prosperity, when Chicago was a leading city in the geography of the Fordist mass-production economy. Manufacturing employment continued its decline on the West Side throughout the 1990s. Instead the neighborhood plays a new role in such postindustrial economic activities as media production and entertainment provision.

North Avenue only recently evinced the worst properties of Scorsese's take on 1970s urbanism, with an active drug trade and garish prostitution. Though these spectacular elements of sidewalk life misrepresent the more ordinary existence led by most neighborhood residents, principally Latino or white ethnic working-class people trying to make the best of constricted economic opportunities, they overwhelmingly colored representations of the neighborhood in the news media. So, although Wicker Park would become in the 1990s a celebrated center of creativity and alternative culture, in the 1980s little news from the neighborhood attracted the attention of the press beyond lurid crime stories, as when a homeless heroin addict was accidentally caught in gang crossfire.[1]

How did such a spectacular turnaround come about? The answer is complicated. In fact, property speculators had already identified the neighborhood as a likely site for gentrification by the early 1980s[2] — as Neil Smith would put it, they already perceived the potential "rent gap" in the neighborhood.[3] That gap was not sufficient to automatically "suck capital back," as Smith's theory would anticipate, at least not yet. But while affluent residents and upscale businesses remained wary of the district, artists were already making a home in the neighborhood during the 1980s, and their numbers increased as the decade wore on. Without discounting the roles of property entrepreneurs and political elites, we will take these individuals seriously as a key force in the neighborhood change that "exploded" in the next decade.

The construction of the Wicker Park scene drew upon both local history and the accumulated mythology of the artist in the city as important resources. Young artists frame elements of the local landscape that many would find alarming as instead being symbolic amenities. Particularly in the nascent stages of the scene's development, moving to Wicker Park meant negotiating the sidewalks with colorful denizens of city streets who alarm suburbanites and who drove DeNiro's taxi driver to psychotic distraction. But artists are committed urbanites, and they fold the representation of neighborhood decay into their picture of authentic urbanism, even as their presence contributes to the reversal of many of its effects. In fact, the afterimages of decay, aestheticized in neo-noir entertainments, heroin chic fashions, or *Taxi Driver* posters, are imprinted on the cultural offerings produced in and through this new bohemia. In this way, the figurative representations of disorder are translated into the beacons of a new symbolic order in the neighborhood: "Welcome to Wicker Park."

Among the residents who initiated the neo-bohemian transition, the disappearing street vice is recalled wistfully. One such resident insists that some of the local artists even engaged in a quixotic quest to protect the right of prostitutes to patrol this stretch of real estate. For new bohemians, Wicker Park is valued for its ability to suggest a gritty edge, to be a neighborhood "dripping bullets laced with sex," as the local alternative magazine *SubNation* put it in the mid 1990s.[4] Christopher Mele describes a similar process in New York's Lower East Side: "While the images and symbols of urban decay remained the same, their representations and attached meanings shifted from fear and repulsion to curiosity and desire."[5]

On the Edge

It is the spring of 2000, and Michael Watson and I sit in the Borderline, a classic Chicago corner bar across the street from the Flat Iron that has been a constant through the neighborhood's transition from the late 1980s. Michael, a thickly muscled African American in his late thirties, worked for several years as a bouncer for this bar and for the Red Dog, the "underground" nightclub that occupies the upstairs. Both businesses, along with a restaurant next door called Café Absinthe, are owned by the same people, and all three venues have reaped the benefits of neighborhood transition by steadfastly expressing neo-bohemian chic, each in its own way. My tape recorder running, Michael and I engage in a long conversation about his colorful history in the neighborhood. A good-natured ex-enforcer, Michael laughs easily while explaining that his job was "to handle the stupid shit": "I was a bouncer here for a number of years. I was a bartender and bouncer at Dreamerz [a bar up the street on Milwaukee, now called Nick's] for a couple of years. I ended up living upstairs from Dreamerz when the owner got shot. I got involved in Wicker Park through violence. Everything at some point had violence."

Despite his imposing physical presence, Michael is no stereotypical urban thug. During the time that he was policing bars, Michael, a poet and writer, was active in the local literary community nurtured in such places as the Guild Complex, the Urbus Orbis café, and the Afrocentric bookstore Lit X. He no longer lives in Wicker Park, but he still is involved in the neighborhood as the poetry director for the Around the Coyote (ATC) art fair.

Michael moved into Wicker Park in 1988, the year before Urbus Orbis opened for business a half block east of the Borderline on North. "Strange enough, what Urbus Orbis was before — there was a big shooting gallery there," Michael says, using the common slang for a site for selling, buying, and injecting heroin. "That's where all the junkies would go. You knew where they were, and considering where it was, this was like, this little walkway, all of this, this is where they would hide." He indicates the stretch between Urbus' former home and the Borderline.

> From there to the bar, it was a [transient] hotel called the Victory Hotel, and they would congregate there. This six corners was always live, and this is where they would go, they would go and hide in the small places. If they could get in there, they would do their shit there, or they would go into the park itself. That was called Needle Park, and you would see hypodermics just out there in the grass.

Given this less-than-bucolic portrait, why did Wicker Park increasingly become attractive to so many young people, often from far more sanitized, middle-class origins? In part, artists' interest in locating in marginal neighborhoods whose majority population is poor and nonwhite involves the desire to occupy inexpensive space adequate to their needs. They are a transient population, breaking ground in marginal urban areas that may be targeted for redevelopment. Christopher Mele notes that, "because of their limited economic resources and/or preferences for residing in alternative neighborhoods, these groups endure above average levels of crime, noise and drug related problems."[6] However, it appears that many Wicker Park residents do not merely tolerate these drawbacks, although there remain limits to the extent that any resident with other options would allow his or her personal safety to be compromised. Participants in Wicker Park's arts community profess an ideological commitment to race and class diversity, although, as we will see, the practical definition of diversity is complicated and often fetishistic. Sharing the streets with working-class and nonwhite residents, even if personal interaction remains superficial, is part of their image of an authentic urban experience.

Uniformly, the young artists whom I interviewed who moved into the neighborhood during the late 1980s recall it in terms similar to those of Michael Watson. Many speak of "gang bangers" congregating around the six corners and, along Division Street on the neighborhood's southern border, members of the West Side Latin Kings.[7] Checking these recollections against empirical indicators supports the depiction of a high-crime neighborhood, and no doubt illicit elements had a disproportionate impact on the street ethos, even if the majority of residents confined themselves to law-abiding activities.[8] Yet, if we ordinarily assume that criminal activity like drug dealing and prostitution would repel most residents or at least be endured grudgingly, the picture for young artists in the neighborhood is more complicated than that. Indeed, the recollections of this period tend toward the nostalgic, as the manifest dangers of the neighborhood coincided with the bohemian disposition to value the drama of living on the edge. As one local entrepreneur puts it, "There's a sense of vitality in the streets. Along with the danger there's a vitality that you lose when you're sure about your personal safety. There's a certain edge that goes away. And there's something exciting about having that edge. There's something [exciting] about having drug dealers right up the street."

Other new residents to the neighborhood in the late 1980s also demonstrated a surprising lack of aversion to those elements of garish street life that standard narratives of urban renewal and community policing identify as social disorder.[9] Nina, a painter who was employed for more than a decade in the local bar scene, also moved into the neighborhood during the late 1980s. "It was mostly Puerto Rican and some Mexicans. A lot of Hispanic families and hookers. And when we moved in, the first people that talked to me were the hookers, and they were pretty nice." She recalls the environment as gritty and dangerous, although her take is somewhat more complicated than simple bravado. Rather than treating danger as a source of allure in itself, she remembers the way that she constructed a sense of safety through the forging of ties with the local community:

After a while people got to know me by sight, and I felt really safe
there. And I remember one time I left my house, and I saw all
these people chasing this one guy down the street, and they started
beating the shit out of him. I thought, "Oh my God." And one guy
came over who lived next door, and he said, "That guy just tried to
steal this girl's purse. I just want you to know that you have nothing
to worry about in this neighborhood, you will always be safe."

This story suggests a different relationship to the neighborhood than
that presented by the entrepreneur who trumpeted the thrill of the edge.
Nina does not simply stride willfully through a landscape of stimulating
dangers. Instead, she locates a sense of community forged through shared
space, offering security for members via practices of informal — and
apparently brutal — social control. Even the prostitutes, committing
victimless crimes, are incorporated into the sense of community; they are
remembered as "pretty nice." For Nina, activities that might suggest to
outsiders the breakdown of social order were in fact governed by deep
norms of community. Further, this community offered her a veil of protection: "I felt safe because people knew me, and although they were mostly
gang bangers, they were also. . . It was a neighborhood and that made me
part of it."
    Unlike many of the artists and hipsters entering the neighborhood,
Nina was born in the city and was already confident about negotiating
a heterogeneous street environment. She is of mixed race background,
which she acknowledges as a potential factor in her acceptance by the local
Hispanic community. "I have the kind of skin tone that I can kind of
pass. Here in the United States, people tend to think I'm Puerto Rican. In
Europe, people think I'm Arabic. I blend in a lot." A graduate of Lincoln
Park High School on the near North Shore, she studied dance for five years
of her adolescence and attended the Art Institute, studying Fine Arts for
several years without taking a degree. In fact, she was enrolled in the Art
Institute during the late 1980s, and she worked as a manager at the Blue
Note, a popular bar with hip young urbanites in the community area of
Bucktown, next door to Wicker Park. Nina's education and involvement in

the arts largely distinguished her from the Hispanic community that was mostly working class, with a high proportion of foreign-born residents. Nonetheless, the combination of being a woman in the heavily male-dominated neighborhood arts scene and being nonwhite may have contributed to her being more readily accepted by the locals, at least provisionally.

Their greater risk of sexual assault no doubt makes women less likely to glamorize the threat of urban violence. Nina did not experience her navigation of the streets as an affirmation of her own toughness; rather, she emphasized the importance of forging provisional bonds with the varying groups within the neighborhood as a key to personal security. On several occasions I was told similar stories by female participants in the arts scene about the young Hispanic men who patrolled the streets offering a "heads-up" on potential dangers and indicating that they would keep an eye out for the woman's safety.

White male informants likewise indicated that over time they would be "recognized" by Hispanic locals and thus were less likely to be hassled, but this arrangement was generally described as implicit, a sign of respect rather than of community per se. Jimmy Garbe, a musician and longtime bartender at the Rainbo Club, recalls, "We coexisted pretty well, I think. In terms of violence, if you kept to yourself you'd feel like you weren't involved in it. I think it scared a lot of people, but for me I'd be walking down the street, hear gunshots, and I'd make sure it wasn't around me, and I'd be alright."

Whatever the level of reported interaction, young Hispanic men enjoyed a peculiar hegemony on the streets, similar to the situation Elijah Anderson describes of young black men in a gentrifying Philadelphia neighborhood,[10] in contrast to their marginal positions in the larger social order.

For male participants in the arts scene, who were mostly white and hailed from relatively privileged backgrounds, navigating the gritty streets involved adopting an "outlaw aesthetic," expressed through both dress and demeanor, that was similar to the persona that Norman Mailer attributed to the "existential hipster" during the 1950s. Such personal styles were intended to mark them as different from mainstream society and to help them blend into the local scene as they experienced it. This persona was not something that they necessarily arrived with; instead, they typically tell a story of an evolutionary process in which hip mastery of the street grows out of the situational context of the local field. The sculptor Alan Gugel, who received an art degree from a rural midwestern university, describes himself as having initially been "like a total outsider from all this. I came here [to Chicago, in the mid-1980s] from South Dakota, so I'm a bumble-fuck; yeah, I grew up on a farm, went to a little college town in Minnesota. I never saw a city of larger than about forty-five hundred people."

He clearly viewed his inexperience as making him inadequate to the challenges posed by the neighborhood street. But that would change.

Alan adopted an aesthetic typical of the 1990s urban hipster — tattoos, secondhand clothes, and stringy, unwashed hair — in order to successfully navigate the uneven neighborhood social terrain.

[Wicker Park] was not a place for folks who in any way look like they're a part of regular society to be hanging out in. And that was one of the things I learned real quickly in Wicker Park, is that if you don't fit in, you're going to have trouble. If you don't look like the greasy slimy artist, you know, you're just going to be accosted left and right. And it happened to me many times when I moved here. I got beat up and pushed around and robbed and everything else. [So I said,] "Okay, I'll turn into a vampire." And it wasn't like I actually made a conscious decision. It was sort of like a gradual process, over a period of time, it was like my whole appearance, from "good little white boy from South Dakota" changed into like "now I can walk down an alley street in the city of Chicago and deal with the scariest people you'd meet there."

The mode of social performance Alan describes is not affected sim-ply to repel danger, though; artists in the scene perform for their peers at least as much as for potential assailants. Adoption of the appropriate demeanor and dress quickly became a means through which insiders were distinguished from outsiders in the art and music scene, especially as gen-trification increasingly eased the sense of real danger. Further, through the circuits of fashion and media, such styles are spread from the gritty urban

milieu, evoking only an enticing *fantasy* of grit and danger in more sanitized locales.[11]

Though women in the neighborhood certainly had their own methods
of distancing themselves from mainstream society, I never heard one describe the navigating of alley streets with the same level of bravado. While
women were a visible minority, and well represented among the success
stories of the arts scene, there is nonetheless a distinct articulation of male
privilege in the normative neo-bohemian construction of the edge.[12] Liz
Phair, who was among the first Wicker Park musicians to achieve widespread critical and popular recognition (*Spin* named her recording debut
its "album of the year" for 1993), has often tweaked the macho pretensions
of the male-dominated neighborhood music scene, for example by titling
her first album "Exile in Guyville."

Nelson Algren's Neighborhood

In the local articulation of hipster sentiments, Nelson Algren often emerges
as the patron saint of neo-bohemian Wicker Park. Winner of the first
National Book Award, Algren is today comparatively obscure in the national literary tradition. But his importance for bohemian aesthetics in general and Wicker Park in particular is significant. "Walk on the Wild Side,"
Lou Reed's classic examination of the decadent urban demimonde, originated in an effort to adapt Algren's book of that name to stage. Moreover,
Algren lived and wrote in Wicker Park, setting some of his best-known
work in the neighborhood. In fact, the city of Chicago, as part of its shrewd
plan to highlight its own cultural depth, has given the honorary designation Nelson Algren Avenue to the stretch of Evergreen Avenue containing
Algren's last Chicago apartment. Algren's other address in the neighborhood,
at 1523 West Wabansia, actually has far more romantic connotations. The
Wabansia address, by all accounts a flat of magnificently bohemian auster-
ity, was not only where Algren penned his best work but also where he was
periodically joined by Simone de Beauvoir in their spectacular transatlantic
affair.[13] Unfortunately for the literary tourist, this site in Wicker Park's
northeast corner "exists only in the imagination," as Greg Holden puts it
in his book on Chicago's literati. "It was bulldozed to make way for the

Kennedy expressway, which cut a swath of destruction through this and other Northwest Side neighborhoods."[14]

Algren wrote his best-respected material during the ten-year period following the Second World War, when Wicker Park was still a working-class Polish community, and he focused his attention on the seamier side of the neighborhood, particularly in the 1949 National Book Award–winning *The Man with the Golden Arm*,[15] which follows the career of a small-time hustler and heroin addict, Frankie Machine (played in the film adaptation by Frank Sinatra). While this period was in many ways the height of Fordist prosperity on a national level, signs of decline showed in this hardscrabble neighborhood, some caused by the incipient pull of the suburbs and others by the decline of the small-scale industrial enterprises that dotted the local landscape, eclipsed by the vertically integrated mega-industries of mature Fordism.

Algren liked to claim that he was turning his literary lens on the "real" Chicago. A former journalist, he saw himself as "the last of the old-style Chicago realists, trapped in the collapse of the industrial order," Carlo Rotella writes.[16] He drew inspiration from the muckraking tradition of novelists like Upton Sinclair, whose most famous novel, *The Jungle*, focused on Chicago's South Side "Back of the Yards" neighborhood and the immigrant workers who lived there at the turn of the century. However, while Sinclair strove to depict the labor conditions of these immigrants (in the process creating a national scandal over the terrible hygiene of the meat-packing industry), Algren, despite his socialist affiliations, showed comparatively little interest in such matters. Indeed, reading *The Man with the Golden Arm* could easily lead one to suspect that no one in the neighborhood held any jobs at all outside of the marginal pursuits of the pool hustler, bookie, pusher, jazz musician, or prostitute.

Thus, despite his ambitions to carry on Chicago's literary tradition with his "naturalistic" fiction, Nelson Algren diverges from Sinclair and others by directing his attention away from industrial labor. This does not imply that Algren was being disingenuous when he claimed that his award-winning novel drew on actual experiences of Chicago's West Side. Individuals like the ones populating Algren's fiction could surely be found in postwar Wicker Park, just as street gangs and prostitutes would be a visible feature of the local landscape during the 1980s. What is important are the consequences of the novelistic choices Algren made as they relate to the artists and hipsters who would make the neighborhood home in the late 1980s and early 1990s.

In his study of the relationship between urban literature and the cities drawn upon for inspiration, Rotella takes Algren's work, especially *Golden Arm*, as a primary source. He argues that "literary writers are in the business of imagining cities," and he elaborates:

> These cities of feeling (to use [Willa] Cather's phrase), which are not imagined from scratch, tend to descend from two sources. One is other texts, since writers read one another and swim in the greater sea of culture, assemble repertoires and influences, repeat and revise. The other source is "cities of fact," material places assembled from brick and steel and stone, inhabited by people of flesh and blood — places where, however sophisticated we might become about undermining the solidity of constructed terms like "real" and "actual' and "fact," it is unwise to play in traffic.[17]

In Algren's work, we can detect not only the influence of Chicago realists, but also continuity with other literary traditions fascinated by the seamy side of urban life. But we can also see the important influences generated by the "city of fact" in its historically specific instantiation.

By depicting highly selective — and not clearly representative — features of the neighborhood as "authentic" Chicago, Algren's work is recognizably within the bohemian tradition that identifies the illicit with the real in the space of the city, and he helps to propel that tradition forward. Bohemians have long drawn inspiration from elements of the urban underworld, aestheticizing the activities of criminals and addicts in their cultural projects. Algren thus participates in the ongoing, collective cultural construction of "the city of feeling" that interacts with the historically shifting "city of fact." This tradition figures in surprising and likely unintended ways in the self-understanding of many artists as they moved into the neighborhood area beginning in the 1980s, applying their constructs in the definition of the neighborhood situation. Frequently,

Algren's legacy would be explicitly addressed in the contemporary constitution of Wicker Park's place identity, and an aestheticized relation to
urban vice would become a key feature of neo-bohemian modalities of
value. Like *Taxi Driver*, *The Man with the Golden Arm* is a cultural product intended as social criticism, but observation in Wicker Park shows
how both works can operate as background for a kind of urban fantasy:
"the city of feeling."

The Bop Shop, opened in the late 1980s, was located across Division
Street from Frankie Machine's fictional address in *The Man with the
Golden Arm*. Though it moved out of the neighborhood in 1995, the Bop
Shop was an important venue in the early development of the neighbor-
hood's hip local scene. While ordinarily it showcased an eclectic mix of
live jazz and spoken word performances, it also featured a celebration of
Algren's work and life annually on the occasion of the author's birthday.
I was present at the Algren showcase hosted by the Bop Shop in 1993.
Throughout the evening, speakers made much note of Algren's unflinch-
ing portrayal of the neighborhood, characterizing his "refusal to sanitize"
neighborhood life as a courageous act of social criticism. The presentations, including the screening of a video production based on Algren's
1947 short story, "The Devil Came Down Division Street,"[18] reproduced
the image of postwar Wicker Park as an exotic slum, and of Algren as its
staccato bard. A local actress portrayed Simone de Beauvoir, hamming
up the feminist intellectual's French accent as she read selections from de
Beauvoir's memoirs recalling Algren's particularly American virility. The
evening's program was titled "Living It Up with Nelson"; fliers for the
event featured Art Shay's well-known photo of Algren in a poker game,
presumably not unlike the games described in his fiction. The image is
of the author as a backroom anthropologist and risk-taker, getting down
and gritty, living it up, then reporting back to fascinated outsiders as a
quasi-native informer.

Steve Pink was among the artists who moved into the neighborhood
in the late 1980s, when he was in his early twenties. Pink was primarily
involved in theatrical production during that period; later he moved to Los
Angeles, where he enjoys success as a screenwriter and movie producer. He
describes how both reading Algren's books and encountering Art Shay's

well-known photo-essay *Nelson Algren's Chicago*[19] helped lead him to the neighborhood:

> I was interested in Wicker Park mainly because I had read [Algren's novel] *Never Come Morning.*[20] . . . I saw [Art Shay's] book. He took the famous pictures of Nelson Algren. He had all these great pictures of Algren walking down Division and stuff. . . . Just the neighborhood seemed so interesting based on that, it just seemed so historically significant, and I think [Algren] was the first winner of the National Book Award . . . so it seemed kind of interesting, kind of a literary figure that I liked because [his work] was about drugs and gangs, and I read about drugs and gangs and hardship and art.

The image of Algren on Division Street evokes long-standing bohemian archetypes: Baudelaire's *flâneur* and Mailer's existential hipster. Here we see an exemplary instance of the interplay between fantasies of the gritty street and the construction of a creative persona in the nascent stages of a young cultural producer's career. Algren's association with the West Side generates an experience of continuity with traditions of cultural creation in Chicago, associations that downtown locales no longer offer. His work adds to the allure of Wicker Park for those who wish to follow in his footsteps, and it contributes to the cumulative texture of the local culture.

Algren's influence on the neighborhood cultural scene was acknowledged by Julie Parsons-Nesbitt, the director of the Guild Complex, a neighborhood nonprofit nurturing literary efforts:

> Nelson Algren was really influential in Chicago writing. He really marked or imprinted Chicago literature in a very particular way. He was a very purely working-class writer. I think that's very much reflected in our work and in the sort of literary stance of Chicago. He wrote about the immigrants who lived in Chicago. He wrote about the poor people, the junkies, the prostitutes. That was a very particular kind of literature that exists and is valued in Chicago and exists and is valued at the Guild Complex. I mean not that we

Figure 4.2 Nelson Algren on Division Street. Photo by Art Shay.

have readings about prostitutes or anything but that what we do reflects people's real lives. The real lives of many different kinds of people. Including people that are often invisible in art, in American culture. And I think that influence has a lot to do with Nelson Algren's work.

The idea that junkies and prostitutes are "often invisible" in art is dubious given the massive canon addressed to these matters. In fact, from Manet's *Olympia* to the opening line of East Village novelist Tama

Janowitz's *Slaves of New York* ("After I became a prostitute I had to deal with penises of every imaginable shape and size," a heroin-addicted prostitute reports),[21] such matters have been repeatedly broached by notorious works of bohemian and neo-bohemian culture. What is important to note is the link between these gritty themes and ideas about urban authenticity — "what we do reflects people's real lives." The owner of a Bookseller's Row in the neighborhood reported that in 1993, the store's best-selling paperback was Bettina Drew's 1989 biography *Nelson Algren: A Life on the Wild Side.*[22]

"The Odd and Charming Refuse of Humanity"

As we can see, the evocation of diversity, both the key value of multiculturalists and the linchpin of much contemporary marketing discourse,[23] is an important part of the ideology expressed by the avatars of Wicker Park's new bohemia. In contrast to theories of the city as trending toward increased homogenization and sanitization in response to the demands of new residents, diversity here is taken to be a central principle of urban authenticity, and the definition of diversity typically proffered by local artists gives special value to the illicit and the bizarre. For an admittedly small but disproportionately influential class of taste makers, elements of the urban experience that are usually considered to be an aesthetic blight become instead symbols of the desire to master an environment characterized by marginality and social instability.

Many of the young artists and aspirants making their way into the scene in its nascent stages came from backgrounds quite different from the world of the streets they would encounter in Wicker Park. Still, they were steeped in the long traditions of the artist in the city, traditions that provided a particular lens with which to interpret their new environment. Delia, a film student at the Art Institute, recalls:

> being a corn-fed midwestern girl walking into Wicker Park, I had never seen a six-way intersection before. But it kind of reminded me of how Greenwich Village looked on TV. Kind of gritty inner city, cars, *homeless people*. So it was pretty much a big culture shock.

I had spent most of my time in either rural kind of pseudo suburban area, or a medium-sized town like Columbus. So there was
nothing like that kind of energy. I loved it. I loved the colors and
the people and the sounds and the streets and the whole bustle.
It was great.

With this account, Delia calls to mind what has become a bohemian
cliché — the wide-eyed novitiate both shocked and seduced by the gritty
glamour of the big city. This narrative animated literary efforts in nineteenth-century Paris that included Balzac's *Lost Illusions* and Flaubert's
Sentimental Education.[24] But now the stranger in a strange land is not a
French lad from the provinces but a striking black woman with a head of
flaring dreadlocks, and her fantasies of the city were nurtured not only by
books but by mass-mediated images — Greenwich Village on television.

Of further interest is Delia's inclusion of the visibly unhoused in her
description of the energy-infused street. The years of gentrification since the
period Delia describes have not eradicated this presence, and it complicates
a general story in which city elites wage war on the homeless population
as part of a comprehensive strategy to render the streets more pleasing to
elite consumers. Regenia Gagnier observes the tendency among develop-
ers and politicians to cast homelessness as an aesthetic rather than a social
problem, with the homeless ineligible for civic consideration because of the
twin crimes of being propertyless and unsightly.[25] When the Viacom-owned
network MTV, the contemporary avatar of hip consumerism, set up shop
in the neighborhood to film its reality TV program *The Real World*, cameras
were reportedly turned off whenever a homeless person entered the frame.

The assumption is that the class of residents and/or viewers that commercial interests wish to attract will hold a uniformly adverse view of these
unwashed street people. Yet Delia tells a different story, one in which homelessness is still an aesthetic principle, but now as part of a street panorama
that she equates with gritty authenticity. Moreover, Delia, a film student
at the Art Institute whose own dramatic persona helped to make the local
scene as she worked in important neo-bohemian venues like Urbus Orbis,
belongs to a category of residents whose aesthetic dispositions cannot be
quickly dismissed as unimportant.

Indeed, though the park is no tent city, there are elements of the neighborhood that contribute to its providing what Mitchell Duneier refers to as a "sustaining habitat" for a limited population of unhoused individuals.[26] Some of these elements are enhanced by the economic development of the neighborhood, since the new economic activity increases the density of pedestrian traffic and thus opportunities to make money for panhandlers and informal vendors. As Duneier notes of Greenwich Village, one of the keys to being a sustaining habitat is that at least some of the local residents have dispositions tolerant of the presence of the truly destitute in their midst. Greenwich Village, pretty much the proto-bohemia of United States cultural history, has become an upscale neighborhood, but the educated professionals who reside there nonetheless often evince socially liberal attitudes toward the homeless, particularly insofar as the homeless maintain a certain informally proscribed decorum. The young artists and lifestyle aesthetes who congregate in Wicker Park likewise are sympathetic to the homeless population, whose visible presence continues to counter the overall trend toward increasingly sanitized streets, and thus helps local artists to maintain their sense of neighborhood diversity.

As Duneier indicates, those who make a life on the streets in the neighborhood need not be simply blight on the landscape, but instead may become integrated into the local fabric and flavor of neighborhood life. They become like the "public characters" Jane Jacobs describes,[27] difficult to displace in a neighborhood that continues to trade on gritty bohemian aesthetics despite ongoing gentrification — at least so long as their behavior remains circumspect and their numbers remain manageable. Resourceful street entrepreneurs attune themselves to the distinct local dynamic. One panhandler in the late 1990s made a habit of studying the *Chicago Reader* so that he could alert nighttime revelers to the various entertainment opportunities surrounding them, as in, "There's poetry at the Mad Bar tonight. And Subterranean has two bands — $5 cover."

This brings us to the late Wesley Willis, the most unusual of the successes to emerge from the widely hailed Wicker Park music scene of the 1990s. Willis, an African American man born in Chicago's public housing, passed away at the age of 40 in 2003. Willis was a diagnosed schizophrenic, intermittently homeless, who in the 1990s came to play an important role in the constitution of the neighborhood as a site of offbeat, funky urban culture. He was unmistakably one of Wicker Park's "public characters." Jim DeRogatis, music critic for the *Chicago Sun-Times*, wrote in 1995, "If you've been to Wicker Park, chances are you've run into Wesley Willis. Willis is a big, burly 31-year-old who's a fixture in Chicago's hippest neighborhood. He walks the streets having agitated conversations with the voices he says he hears in his head."[28]

One of the few individuals from the neighborhood who might reasonably bear the tag "outsider artist," Willis also engaged tangible people on the sidewalks with his push to sell his drawings and CDs, a contemporary version of Greenwich Village's indigent poet Maxwell Bodenheim in the 1940s.[29]

In a different neighborhood, Willis might have languished as an unfortunate casualty of nationwide deinstitutionalization; in Wicker Park, he became a local and then a national celebrity who would tour the country with a backup band of genuine musicians, a group named the Wesley Willis Fiasco, performing his bizarre music. DeRogatis provides this description: "As the group plays powerful, free flowing grooves, Willis takes to the stage to rave about his favorite bands (everyone from Radiohead to Sabalon Glitz) and rant about the things that tick him off (a man who assaulted his mother, the bar closing early, or Casper the Friendly Ghost)."[30]

Though music critics have for the most part discerned little actual talent on display in these performances, Willis numbers among his fans and promoters the local band Urge Overkill and nationally successful musicians including the Beastie Boys, Henry Rollins, and the Red Hot Chili Peppers. With the help of such high-profile support, Willis performed in well-known venues around the country, cut a two-album deal with American Records, and appeared on MTV. When not on tour or hospitalized, Willis continued to spend his time raving and vending on Wicker Park's sidewalks.

Like many other features of the neo-bohemian scene in Wicker Park, the cult following that Willis achieved has bohemian precedent. Jean-Michel Basquiat's teenage bouts of homelessness in Washington Square Park are a part of the late East Village artist's enduring legend.[31] Even more to the point is the fame afforded to Joe Gould, the Greenwich Village street person sometimes referred to as the "last bohemian," immortalized in Joseph Mitchell's classic essay "Joe Gould's Secret."[32] Ross Wetzsteon writes of Gould, "He transformed vagrancy into an ideal. . . . Joe Gould wasn't just a bum, he was a bum of a certain genius. He was the leading beneficiary of the Villager's enduring fantasy of a link between the social misfit and the cultural rebel — after all, who better understood society's hypocrisies than society's outcasts?[33]

A half century after Gould prowled the streets of the Village, Willis would animate similar fantasies among the hip in Chicago and around the country.

Some critics expressed concern over the exploitative nature of turning a mentally ill street person into a rock 'n' roll novelty act. Buddy Seigal wrote in the *Los Angeles Times,* "In trumpeting Willis, [rock musicians] seem to be laughing not so much with him as at him, in a 'let's hang out with a retard and have a giggle' display of rank insensitivity, mindless pack mentality and rampant egotism."[34] But as Libby Copeland notes in a *Washington Post* feature on Willis, for others "he's the ultimate punk. He says what he wants. He's so . . . authentic!"[35] Erik Wulkowicz, a long-time participant in the local arts scene, argues that Willis' appeal lay in his authenticity: "Here's a guy that just because he's crazy and absolutely real has entertainment value. Some people for a good reason want to do something with him. The rest of them suspect that they are trying to cash in on the guy." Meanwhile, Willis seemed to genuinely enjoy performing and clearly benefited materially, as his minor celebrity showed surprising longevity. The money he earned rescued him from the streets, and his band mates doubled as caretakers on the road: "They forward messages to him, remind him to bathe," Copeland wrote.[36] Things could have been worse, and, of course, for many who are similarly afflicted, things are worse.

The celebration of Wesley Willis by hip tastemakers runs counter to the elite strategies of concealment and containment that Mike Davis demonstrates are typically directed against the homeless, some of whom are mentally ill, in Los Angeles.[37] On the other hand, the spotlight shone on this "outsider" also obscures the legions of uninvited schizophrenic individuals trapped in destitution and desperation. They occupy quasi-carceral street environments and the overcrowded prisons of their own minds, often doing harm to themselves and sometimes, dramatically, to others. The revanchist policies enacted by civic elites around the country against such individuals show that those individuals lack rights in the city. How can they be citizens in a neoliberal landscape where participating in the consumer cornucopia is the mark of legitimate belonging?

Willis, on the other hand, is granted citizenship in the world of hipster culture, but it is what Lauren Berlant might call an "infantile citizenship," secured by the schizophrenia that renders him apparently transparent and therefore "real" for his audience. Willis embodies nostalgia for jaded culture-industry veterans like Rollins or the Beastie Boys for a time when bohemia could be experienced as simple utopia, the fantasy of the rock 'n' roll lifestyle. Licensed by a mental illness that made him "like a child" and therefore free to articulate any sentiment, Willis could happily declare desires that bohemians typically let go unspoken: "I want to be famous and rich." Meanwhile, a young neighborhood Internet entrepreneur whom I interviewed was effusive about Willis as evidence of the neighborhood's offbeat character, and, yes, diversity.

Heroin Chic

The relationship to the exotic on Wicker Park streets, whether represented by Willis' schizophrenia or by the presence of gang members, prostitutes, and users of illegal drugs, was not limited to cool voyeurism. For Alan, the industrial sculptor and erstwhile self-described "bumblefuck" from South Dakota, development of a heroin habit of his own sped the acquisition of the appropriate dispositions for negotiating the terrain and becoming in some sense an insider. Indeed, drug use was one way that newcomers would straddle multiple worlds in the streets. The norms of this complex street environment, in which definitions of hipness became intertwined with the "underworld," are embodied in the dispositions of participants — the drug habit bleeds into habitus, with this new style not necessarily experienced as a "conscious decision."

The Wicker Park drug scene captured many participants who were also active in the arts, and it had an impact on the articulation of style on Wicker Park streets even for non-users. This aesthetic made its way into the mass fashion market during the 1990s in the form of "heroin chic," valorizing forms of the street found in locales like Wicker Park and New York's East Village. Michael Watson confirms that "every now and then the two worlds [of artists and drug dealers] merged. There were a few artists I know that had to leave the neighborhood because they got involved in the drug aspect. . . . You'd always know that so-and-so was in rehab." Alan himself successfully kicked his habit, and has gone on to commercial and critical success as an artist.

Drug use in the neighborhood does not merely reflect reckless hedonism. Experimentation with narcotics is also part of the bohemian tradition, a well-known feature of the biography of many admired cultural producers. In a *New York Times* article marking the death of Layne Staley, lead singer of the Seattle-based grunge rock band Alice in Chains, John Pareles writes, "A Romantic ideology that predates rock glorifies the self-destructive artist as one who's too honest and delicate for this world. As the myth goes, artists use drugs or alcohol to free up inspiration and insulate their sensitive souls from ordinary life."[38]

Continually informed and updated by songs like the Velvet Underground's "Heroin" and "Waiting for the Man," or books like William S. Burroughs' *Naked Lunch* and *Junky* and Jim Carroll's *The Basketball Diaries,*[39] a mythos surrounding drug abuse forms among the bohemian urban intelligentsia, and in Wicker Park, track marks were available for interpretation as markers of bohemian authenticity. Though drug use is stigmatized in many social venues, occasionally it seemed to have the opposite effect in Wicker Park. Daniel, a young denizen of the scene who did not use hard drugs, recalls, "When I moved into the neighborhood, I was told, 'You've got a good look for this neighborhood. Skinny, the stringy hair, the beard — you look like you *might* do heroin.'" In another interaction I witnessed, a young woman was being advised against romantically pursuing a local artist who had a well-known heroin habit. "But he's so cool," she protested.

For the young artists in Wicker Park, the effect of heroin is not merely sensual; it conditions aesthetic appearance and nurtures a profoundly blasé outlook, the very epitome of a "cool" disposition. One Wicker Park artist and recovering addict suggests that sensual and aesthetic effects dovetail in the scene: "It's part of the idea of being relaxed and enjoying yourself and being willing to participate in what might be victimless crimes for the sake of *aesthetics* or entertainment." Still, the drawbacks of heroin use are also well known, and most people, including most Wicker Park residents, resist heavy involvement. But the "look" associated with drug use, like other aesthetic principles, can be disembedded from the practice. Heroin chic represented the crossover of this aesthetic to mass consumption. Model Zoë Fleischaeur seemed to blame her own addiction on the aesthetic principles dominating the New York fashion world in the mid-1990s: "They wanted models that looked like junkies. The more skinny and fucked up you look, the more everyone thinks you're fabulous."[40]

Identification

As these examples demonstrate, the complex of styles and symbols that coalesce in the construction of Wicker Park's new bohemian scene draws upon the cumulative influence of past bohemian districts. More is at work than just mimesis, however; neo-bohemia is not, as some might suspect, simply a shallow caricature of bohemias past, just another urban theme park. Individuals living real lives with genuine commitments do actual creative work in Wicker Park. Rather than mere homage, the freighted embeddedness of Wicker Park's neo-bohemian scene in the traditions of past bohemias is crucial to its advantage as a site for the active, ongoing production of culture. Participation in the scene can be a crucial part of developing a creative persona for participants well acquainted with these traditions, identifying with city streets of the sort that nourished the imaginations of so many creative predecessors. By entering into the local scene of artists and hipsters, individuals signal their artistic commitment to both themselves and others. They identify with the neighborhood aesthetic and the local arts scene through residence and participation, and, as Andreas Glaeser points out, "identifications are the building blocks of the hermeneutic process called identity formation."[41]

Remembers Alan Gugel: "From the point of moving into Wicker Park, I *became* a Chicago artist." This statement is not trivial; adoption of this persona is a precursor to a range of actions that would otherwise be unlikely. Culture work is filled with uncertainty and disappointment; aspirants face both financial and identity risks in the pursuit of their vocations. Identification with bohemia's traditions of the edge helps sustain necessary levels of commitment in the face of this reality. It provides a model that incorporates the possibility of failure, at least in the short term. Thus the neighborhood does not just magnetize creative talent; it also nurtures crucial dispositions. In neo-bohemia these dispositions, while often explicitly cast against the "mainstream," in fact end up being useful to a host of new strategies for urban development and commercial enterprise.

Because young adults in the arts can rarely compete in terms of income with professional peers, the performance of cultural distinction becomes all the more crucial to their sense of status. But for young people in contemporary consumer society, this performance is complicated by the broad dissemination of subcultural styles across social strata. Herbert Gans notes the important distinction between total and partial cultures in the youth scene.[42] Some young people attempt to create a total culture of stylistic expression, getting a tattoo, joining a band, using unsettling drugs, and generally enacting a stance of mainstream repudiation. This repudiation is challenged, however, by the much larger partial culture of dabblers who may sporadically engage in some of those things while mostly being content to go to school and hold ordinary jobs. Since the division between these two categories can be blurry, participants can reassure themselves of their authentic commitment through spatial identifications that less committed individuals would be reluctant to make.

Given their desire to associate with the "fringe," while still having access to galleries, good bars, and school at the Art Institute, it is not surprising that newcomers to Wicker Park soon resented those that followed and upset the ecological balance. By 1994, such residents were far less quick to give newcomers the benefit of the doubt. A *Chicago Reader* article cataloguing growing anti-gentrification sentiment, titled "The Panic in Wicker Park,"[43] makes clear that the most noisily panicked were usually residents who had themselves been there for only a handful of years at most. Given that their own presence was heavily implicated in neighborhood change, they may have been enacting a version of what

Rosaldo calls *imperialist nostalgia*, "where people mourn the passing of what they themselves have transformed."[44]

As John Irwin has noted with reference to past urban subcultures, the popularity of the scene can also lead to the compromise of its "carrying capacity."[45] Policing of boundaries between insiders and outsiders in the neighborhood bars and other public spaces took on a nastier tone as Wicker Park's bohemian visibility increased. The painter Tom Billings — who has since achieved national recognition, including lucrative corporate commissions — was particularly known for making things uncomfortable for newcomers. "He would just make sure you *really* wanted to be there," laughs one local artist. But like the lines outside Studio 54 during the 1970s, this exclusionary aspect also increases the mystique and desirability of the scene. Moreover, the ability of former stalwarts to maintain boundaries in this fashion came under increasing assault as the decade wore on.

In a revealing episode during the mid-1990s, I watched two young hipsters play pool at the Bop Shop on an evening when a meeting of the local "neighborhood watch" convened in the adjacent room. One youth, barely twenty and decorated with tattoos and piercings, was noticeably alarmed. "It's the Wicker Park neighborhood watch," he said with fierce animosity. "It's like Chemlawn. People in the suburbs put Chemlawn on their lawns to make the weeds go away. Here it's Chem-corner. They want to keep the corners *crispy clean*." This landscaping process would cleanse the neighborhood of its glamorous grit and undermine his flight from suburban homogeneity.

For to be on "the edge," with all the valences that attach to this term, is crucial to neo-bohemian identification. As one West Side gallery owner put it, echoing a familiar bon mot, "If you're not living on the edge, you're taking up too much space." This space of the edge is narrow, resists crowds, and entails a precarious balancing act. To slip too far to one side or the other is to lose it. Through identification with the gritty neighborhood, participants reassure themselves of their legitimate claim to edginess, though the neighborhood itself also teeters uneasily on the verge. Historically emergent themes of bohemia inflect the experience of the local street, in which a range of normative associations take shape: hipness, intensity, diversity, authenticity.

In a 2002 article in the *New York Times* travel section, significantly titled "The Many Accents of Wicker Park," neighborhood resident Brenda Fowler locates the appeal of the neighborhood in its eclecticism:

> Like me, [tourists] probably like the view on the ride out from the Loop on the elevated tracks; the bustle of a working neighborhood; the dingy old bars and the funky new ones; the 24-hour restaurants and those with pedigreed chefs. They like the used bookstores, the dusty old shops and the sparkling trendy ones. The neighborhood feels "alternative," but also trendy, though many people who live here would be repelled by such categories.[46]

If increasingly detached from the reality of more upscale residence and commerce, gritty accents remain a feature of neighborhood character. Moreover, the gritty aesthetic of the local scene, read through the historically constituted lens of bohemian tradition, is imprinted on the aesthetic representations produced by cultural creators, representations that evoke the glamour of urban instability, available to be consumed at a safe distance.

CHAPTER 5

Living Like an Artist

I really love your hairdo, yeah
I'm glad you like mine too
See what looking pretty cool will get you
. . . But if you dig on vegan food
Well come over to my work
I'll have them cook you something that you'll really love
Cause I like you
Yeah I like you
And I'm feelin' so bohemian like you.
    —The Dandy Warhols, "Bohemian Like You," 2000

ALAN GUGEL RECALLS OF HIS EARLY YEARS in Wicker Park: "When I moved into this neighborhood, that's the one thing they kept saying, 'There's a lot of artists, a lot of artists, move down there.' And when I got there, there were no arts. I couldn't see them." Word of mouth indicated that Wicker Park was a likely site for a penniless young artist to find cheap rents and like-minded individuals, but those who moved in during the 1980s at first found that the local scene was comparatively underdeveloped, lacking public spaces to which newcomers might easily find their way. In the late 1980s, Wicker Park was a neighborhood on the verge, but it required the increasing preponderance of commercial spaces geared toward the needs of the artists and aesthetes moving there for the scene to become recognizable to both insiders

Figure 5.1 New Development in Wicker Park.

and outsiders, and for the neighborhood to finally veer onto the path that would differentiate it as Chicago's 1990s bohemia.

Bars, restaurants, and coffee shops were crucial to Wicker Park's emerging neo-bohemian scene, dramatically elevating its visibility to insiders and outsiders alike. They belong to the category of social institution that Ray Oldenberg identifies as "third places," helping people to make new social contacts and thus extend the local community.[1] Oldenberg is inspired in part by the tradition of café society evident in classic bohemian districts, and indeed sites such as Montmartre, Greenwich Village, North Beach, and the East Village all featured well-known venues — Le Chat Noir, City Lights Books, Café Trieste, the Mudd Club, and a host of others — that helped build the bohemian congregation. Such local institutions both drive neighborhood identity and reflect it. Bars and restaurants were hardly unknown to Wicker Park before its neo-bohemian turn, but the character of local venues would change as they increasingly served a new population with distinct social and aesthetic dispositions.

The legacy of the past stages of neighborhood life and demography is not swept away. They form the inheritance of new residents, both in terms of the built environment and the cultural palimpsest. But local businesses that had once catered primarily to members of the Polish or Latino working class would suddenly see a surge in patronage by tattooed young urban hipsters. Local artists — many of whom, like Alan Gugel and Erik Wulkowicz, had advanced carpentry skills — accelerated the rehabilitation process, trading labor for reduced rents or working as local contractors. Michael Warr, the founder and former director of the Guild Complex, points out how the mix of talents contributed to the reproduction of the material landscape, now geared to new styles of culture and commerce:

> When you think of all the performances that have taken place just in Wicker Park alone and all the types of work that go into building stages, you're talking about a lot of skills. Take some of the cafés that have gone up. If you look at Earwax, for instance . . . most of the work that was done in that café was done by artists in the neighborhood. One of the things that was great about running the Guild Complex in Wicker Park was that you never had to go far to find what you needed to get things done in terms of design, in terms of printing, in terms of the artists themselves for programming.

The presence of venues like the Guild Complex, a literary nonprofit that has occupied numerous neighborhood locations, increases the attractiveness of the neighborhood for new waves of artists, and the growing number of artists increases the attractiveness of the neighborhood for further investment. Steve Pink tells of the work his theater company did rehabbing space in the Chopin Theater, which shares the building that currently houses the Guild Complex:

> We opened the theater. It was owned by a Polish immigrant named Ziggy, who's still there. We approached him because he had this great space in the Chopin Theater. . . . We opened that space. We made a good deal with him because our theater company was doing pretty well by that point. So we were like, "OK, you don't charge

us that much of anything for the entire space, and we will build out the space, we'll build everything. We'll build out your space entirely, we'll put in a grid, we'll build in the risers, we'll build the stage." He was going to outfit the front for a coffee shop, anyway. "You do all that, you don't charge us rent on the theater, and we will make sure it's sold out. Don't worry." We did *Fear and Loathing in Las Vegas*, that was the first play we did there.

As this shows, artists in Wicker Park help "make the scene" not simply by providing local color, but also with real brow-sweat.

Phyllis' Musical Inn, my own first point of entry into the neighborhood, was among the earliest venues to capitalize on the coalescence of young cultural creators in the neighborhood, and in the 1980s it helped to make the emergent scene visible, thus further magnetizing participation. One future entrepreneur recalls that visiting Phyllis' helped to alert him to the neighborhood as a potential investment location.

Phyllis' was my introduction to Wicker Park. . . . I had a friend who dragged me down here several times to see bands. That was back when Phyllis' was fundamentally the only thing — we're talking ten years ago [mid-1980s] — Phyllis' and the Gold Star [a bar located across the street on Division] were the only places where non-ethnic white people could go without being in serious jeopardy of their lives.

The grittiness of local bars like Phyllis', which only recently had been propped up by a hardscrabble clientele of grizzled working-class boozers, fit in well with the general construction of neo-bohemian authenticity in the neighborhood, and several such venues were co-opted into the hipster revolution in the manner of Phyllis'. Local performer Shappy Seaholtz recalls of the early 1990s: "There were some bars, but not the kind of bars you want to hang out at except for maybe the brave few that would go to Phyllis' or the Gold Star or the Rainbo. Even then, it was full of hipsters, and I went to a lot of cool art openings and shows." For the most part these venues were not yet widely recognized outside the neighborhood as potential entertainment destinations, and early stalwarts recall them with typical nostalgia:

> Are you familiar with the Borderline? Do you remember what the Borderline looked like in 1989? It used to be you would walk into this place that's literally like a wall, with a bunch of stools on it, with a bunch of cigarettes on one side and a bunch of beer on the other side, and it was covered with this real thick Plexiglas, and it was probably the most dangerous place to be in. About the only places that were relatively fun was Dreamerz and the Rainbo Club. . . . Dreamerz was a total biker bar — that was a rough kind of area to go into. I stayed away from that place on Friday and Saturday nights, I'd go down to Rainbo because that was a little bit more relaxed.

The Rainbo Club, located at Damen Avenue and Division Street on the southern fringe of the neighborhood, became one of the first bars in which local artists and musicians would congregate. A recognizable scene began to take shape, and many musicians whose crossover to national success would help fuel the neighborhood's celebrity were among the regulars, including Phair and the members of the band Urge Overkill (whose cover of Neil Diamond's "[Girl] You'll be a Woman Soon" was featured on the soundtrack of the hipster film sensation *Pulp Fiction*). Jimmy Garbe, a musician who has worked behind the bar at the Rainbo for more than a decade, remembers the scene as not particularly hospitable to outsiders:

> We were kinda pegged as elitist, and I think it's because people would come here, and there would be six or seven different art conversations going on, or possibly music conversations. It was a kind of social (event) for artists to talk about art and music. Some people would come in here, and they'd kind of dig it, but they wouldn't last more than a couple of weeks. They would feel like outsiders, and they wouldn't last, and they would stop coming in. On the other hand, if you were into that scene, it was definitely the place

to come because you could walk in and jump in a conversation at a moment's notice if you were into that.

In addition to providing places for hanging out, new spaces in the neighborhood heightened the visibility of local creative efforts. Many new art galleries were opening in the nearby River West area and also in Wicker Park itself. In 1988, the Ricky Renier Gallery opened in what the *Chicago Tribune* still described as "the wilds of Wicker Park,"[2] and quickly switched from displaying European imports to showcasing local talent. In a practice similar to that of the East Village, local bars began to display the work of neighborhood artists, and they still do. Venues showcasing the neighborhood music scene proliferated. On the same block as Phyllis' Musical Inn, live music could be heard at the Czar Bar and the Bop Shop, both venues whose offerings evolved with the changing neighborhood dynamic. As the neighborhood's celebrity increased, the Double Door, a much larger musical performance space, opened. In addition to local acts, it showcased national fare, including such famous rock acts as the Rolling Stones and Smashing Pumpkins.

But even as the neighborhood became more popular and more expensive, the local aesthetic continued to display the image of grit as glamour. This image connoted an authentic bohemianism appealing not only to committed participants but also to sporadic consumers. Michael Warr describes the Hothouse, a performance venue that has since moved from Milwaukee Avenue to the South Loop as having a "real kind of bohemian feel. . . . That it wasn't just a bright shiny new thing, and when I see bohemia, that's what I see. There's a little bit of the Old World, as close as you can get in the United States. Bohemian isn't just the look, but I do think for me, that kind of grit, I think that is relevant to the picture of bohemia."

Particularly with the popularity of the bare-bones rock 'n' roll styles identified by labels as "grunge," "alternative," and "indie," such venues satisfied the expectations of a range of consumers, and by the 1990s, patrons were drawn from the far reaches of the Chicago area.

In her ethnographic study of "alternative hard rockers" in Wicker Park, Mimi Schippers illustrates the disdain that many committed scene members expressed when increasing numbers began to invade the gritty venues of the neighborhood:

> We pass the Empty Bottle on our right. The Empty Bottle is a local rock club [on the western fringe of Wicker Park] that serves up strong, cheap drinks and features both local and touring bands for relatively low covers. It is one of the scenes that Maddie and other active participants in the alternative rock scene frequent. However, tonight Veruca Salt, a local band who made it big, is playing to a sold out crowd. Maddie looks at the people waiting outside to get in and grimaces. "What a nightmare. Can you imagine being in there? It's probably packed and full of 708ers. . . . People from the suburbs. The area code. All those assholes who wouldn't be caught dead in this neighborhood three years ago, but now because Wicker Park is the hip spot, they flock here. Especially for, like, Veruca Salt. Fucking MTV."[3]

The fact that Schippers' informant Maddie was herself a product of the suburbs is a common inconsistency among scene participants, as they make distinctions that place themselves firmly on the cool side of the fence. A species distinction is maintained in noting that these presumptive suburbanites would not have been caught dead in Wicker Park until recently, and anyway could not have found their way there without a heads up from the agents of the corporate mass media, like MTV or the *New York Times*. In 2001, some locals took the opportunity to protest directly the influence of MTV on the new urban bohemia, attempting to disrupt MTV's staging of its program *The Real World* in a local loft.

The protests against *The Real World* were prefigured by the negative neighborhood reactions to the effects of its showcase arts festival. The ATC was launched in the late 1980s by Jim Happy-Delpeche, a French expatriate art dealer. Its intention was to showcase the work of local artists, with displays in local business venues and in the lofts of artists themselves. Of course, it also had the effect of illuminating the neighborhood's ongoing redevelopment to a wider audience, changing the image of the neighborhood from urban "wilderness" to hip cultural destination.[4] Though the festival clearly abetted the careers of several artists, it also became a center of neighborhood controversy, spawning significant opposition and creating strong cleavages within the community.[5]

The controversy over the ATC shows the contradictions that artists face in the convergence of ideological preference and practical considerations. While many members of the arts community had an ideological commitment to maintaining the neighborhood ethos and resisting further gentrification, they also want to sell their work. Thus, few recognized neighborhood artists actually boycotted participation in the festival. However, for many participation was reluctant and surly. Meanwhile, the "Lumpens," the local subgroup behind the leftist neighborhood magazine *Lumpen Times*, were widely believed to be behind (at least spiritually) the defacement of ATC fliers and posters and the dissemination of literature suggesting possible "guerrilla" tactics to disrupt the festival. Despite local opposition, the festival persists and is gaining popularity. It has been crucial to the growth of Wicker Park's citywide reputation as an arts center, and to its concomitant redevelopment as an entertainment zone. As it became increasingly established, the ATC enjoyed significant corporate sponsorship and the support of local businesses.[6]

Urbus Orbis 1989–1998

Scene stalwart Sid Feldman recalls that the artists living in the neighborhood in the late 1980s "spent a huge percentage of their income on alcohol, or their drug of choice." The figure of the boozy artist is legendary, but another beverage has equally strong associations with bohemia. Coffee would replace sparkling water as the signature beverage for supercharged yuppies in the 1990s, but for artists it has played a central role since the Parisian prototype of bohemia. Seigel points out that none other than Henri Murger wrote late into the night animated by copious ingestion of caffeine.[7] In the U.S. tradition of bohemian literature, the signature work is undoubtedly Kerouac's *On the Road*, whose key draft was "the product of a three-week typing marathon said to have been stoked by Benzedrine and coffee." While Allen Ginsberg liked to claim that Benzedrine sped Kerouac through production, Kerouac himself denied this. "I wrote that book on coffee. . . . Benny, tea, anything — I know none as good as coffee for real mental power kicks."[8] Herbert Gold's rhapsodic memoir of U.S. bohemia identifies it as "where art, angst, love, and strong coffee meet."[9] Even more important than fueling solitary binges of creative work, the coffee house, whether in Montmartre, *fin-de-siècle* Vienna, or Greenwich Village, provided a site for the long dalliances of bohemia, where ideas, half-baked or otherwise, mingled through countless refills.

Sophie's Busy Bee, a famed greasy spoon presided over by Sophie Madej, a stern Polish matriarch, was located just off the six corners. It was extremely popular for its hearty and reasonably priced plates of pirogi, particularly given the often-cramped budgets of aspiring artists. Moreover, the ambiance screamed authenticity; it was a legitimate working-class joint, populated by grizzled old men at the counters and heavyset, middle-aged waitresses impatient for one's order. But the no-frills, working-class Busy Bee was markedly inadequate to other requirements of the young people aspiring to the café society of past bohemias. Alan recalls, "I was going to the Busy Bee for coffee and to read my book, that sort of thing, and they kept throwing me out, they kept saying, 'Order something or go away.' So they didn't want me hanging out in a booth or spending too much time there." Likewise, Michael Watson indicates, "You had the Busy Bee, you got to get in, and you gotta go. It's not conducive. You can't sit there nursing a cup of coffee for five or six hours. It's a working-class place, you don't do that." Thus, while the Busy Bee was much beloved (by me as well) and its passing in the middle 1990s duly mourned, it was poorly suited to the idleness and long conversations that are the stuff of bohemian café society and could not become a center for the construction of a thick local scene of hipsters. The absence of such a meeting place inhibited the formation of the local scene, a barrier effectively lifted in 1989 with the opening of the Urbus Orbis Café.

Urbus Orbis was opened by Tom Handley in a converted warehouse on North Avenue, and it quickly became a popular hangout for artists, musicians, and other young hipsters. The café explicitly catered to "the tattooed, the body pierced, the philosophically open minded,"[10] that is, to the emerging "alternative" scene. As we saw in chapter 1, the building itself reflects the neighborhood's transitions through its various historical uses, having been a dressmaker's sweatshop during the neighborhood's industrial heyday and reputedly a shooting gallery during its postindustrial decline. The transition to superhip coffee shop coincided with an overall turn in neighborhood fortunes. As gentrification of the neighborhood ensued through the 1990s, the space upstairs from the café became an outlet for futons and handmade furniture, an interesting contrast to the discount furniture outlets that have lined the commercial strip a block away on Milwaukee Avenue for decades.

Alan Gugel's recollections show just how intense the pent-up demand was for artists, and also instructs us on the way that the creative talents and manual labors of such individuals shaped the conversions within the neighborhood. "I remember coming home one night after being thrown out [of the Busy Bee], and I saw Tom's sign up in the window that said 'Urbus Orbis, coffeehouse/bookstore opening soon.' So I saw the light on, and I went up there and said, 'Look, can I volunteer some time and help you guys open sooner?' Because I was dying for a place to hang out."

Alan, whose artistic medium is industrial sculpture, contributed more than just ordinary carpentry to the finished product of the café's interior. He designed and built the tables protruding from the brick walls, ornamented by metal gears and other bits of machine detritus embedded in their lacquered finish. Handley, the owner, instituted a practice that would be mimicked throughout the neighborhood, rotating the work of local artists as decoration for the walls and listing prices for sale alongside them. Indeed, it was the display of such art that motivated Happy-Delpeche to conceive and launch ATC.

The nascent arts community responded to Urbus Orbis' opening with immediate patronage. "Our initial clientele was definitely out of the art community," Handley told me. "There wasn't much of a gentrification scene at the time . . . so the art community was my first prop." Unfortunately — and this would be an ongoing problem for the coffee shop's commercial viability — this patronage often consisted of interminable loitering over a single cup of coffee. "There are few places where it's acceptable to table hog," the *Chicago Tribune* reported in 1993, "and Urbus Orbis is one of them."[11] It provided a site for the long hours of "just hanging out" that artists and their friends like to spend while awaiting the lightning bolt of inspiration. Says Michael Watson:

> If you went to Urbus Orbis — and in those days I would spend so much time there — Urbus was such an institution at that point because you could literally sit at a table, you get into a conversation, and next thing you know, you're involved in six conversations. You come in at 2 in the afternoon, and you look up it's 9. You've blown seven hours. You don't know exactly what happened, but you've had all these cool conversations. . . . We'd all sort of end up at Urbus and sit there, nursing a bottomless cup of coffee for $1.35.

Thus, Urbus Orbis allowed for interactions that fostered a sense of community, and opportunities for mutual support and collaboration among aspiring cultural producers. According to Watson, "It was a community. More so than it is now because you couldn't sit there [then] and have no one know what you did. Everyone was doing some sort of art project or political project, or something like it."

On the other hand, lest the utopian recollections carry us away, it is important to remember that communities tend to constitute themselves as much through exclusion as through inclusion. Counter to Oldenberg's enthusiastic portrayal of third places as social levelers, Urbus Orbis was a stage where subcultural distinctions were performed via the snubbing of outsiders. Heavily invested in their status as superhip cultural creators, regulars and staff were less than welcoming to individuals who did not fit their definitions of desirability. Generally this meant those unfamiliar faces whose dress violated hipster norms in the direction of being too upscale and conservative. These individuals were derided with the generic label of yuppies. As Jenna Hill, former manager at Urbus Orbis, recalls, "Often there, if you're not a regular, and if the wait staff doesn't like you, then it's 'fuck you.' That's basically their attitude." Given the standard image of the yuppie as one that hemorrhages his or her considerable disposable income at every opportunity,[12] the practice of making them uncomfortable at Urbus Orbis was probably not sound business strategy. In any event, it demonstrates a common bohemian paradox: elitism is performed through the snubbing of presumptive elites.

Although much greater social distance in terms of education and cultural dispositions separates Wicker Park artists from their working-class neighbors than from young professionals, it would confound artists' commitment to diversity to construct the former as the "other" in the neo-bohemian classificatory system. But few among the Latinos and white ethnics who made up the neighborhood's majority population patronized the café regularly. Some younger people, children of this working-class population, did find their way into venues like Urbus Orbis as regular customers, and for some of them the neo-bohemian scene provided an alternative conception of their life chances than that provided by the bleak landscape of postindustrial poverty and crime into which they were born.

Raul is one of these. Raul's parents are Mexican immigrants, and his older brother was deported on drug charges. He recalls of his high school experience during the 1980s: "At my high school, there was only so much you could do. And I see the outcome of it. I see people from school, and we talk, and all the girls are pregnant or they have kids. . . . And most of the guys are either dead or in jail. Gang banging." Raul found an alternative paradigm in the local arts scene. After discovering the work of Pablo Neruda in high school, he developed an interest in poetry. His tough-kid background helped him gain employment as a bouncer at the Borderline (where he worked with Michael Watson). Contacts there led him to Urbus Orbis, where he would expand on his autodidactic literary interests in interactions with other artists at the café.

> I would spend so much time at Urbus Orbis — not much money, but so much time. . . . And I loved that place because they treated me very nicely there. Every employee that would come through that place met me, they were introduced to me, and they would say, "That's Raul, don't worry about him, he's a nice guy." It got to the point where I was walking around the counter to get my own drinks. I ended up helping my friends when it was really busy and slammed, and not expect anything in return because I spent so much time there. I would take a space at a table, with a cup of

coffee, bottomless, forever. Throughout the years, I met a lot of great people there, writers, artists. It was a good congregating spot.

Now bartending at Borderline, Raul has earned much local prestige (if little income) from his writing skills, and he has stayed clear of the troubles that led to his brother's deportation. Still, as we will see in chapter 6, employment in the local service sector is not an unambiguously happy ending for Raul or the countless other artists in that position, as these jobs carry with them considerable drawbacks.

Suzie Spoliwka, a daughter of Ukrainian immigrants on the West Side, frequented Urbus Orbis during her high school years. She credits the café with leading her away from the activities of her neighborhood peers that she viewed as a dead end.

[The neighborhood] was really rough, you know. Like my friend, she was a lot older than me. We always drove around in her car. She's got four kids now, and she still lives with her Hispanic boyfriend that used to be a Latin King [gang member]. Like a couple of blocks away from here. That's kind of what happened to those people that never left and got sucked up in this lifestyle. . . . And then I discovered Urbus, and I was there all the time. . . . It was a place I could go and smoke and drink coffee. I ended up adopting that place. People would stop by and hang out with me there; I'd get phone calls there. It was an escape for me, because it wasn't the normal direction that people from my neighborhood would be going. . . . For me it was just like normal people [at Urbus]. They weren't pregnant or my family. I discovered that there was more to life than my neighborhood, and I was going to be leaving to go to school. It was when I started to discover the city. I just felt really comfortable there. In my neighborhood I was the freak. And I'm not a freak.

Low-income ethnic communities in large cities are often quite parochial, as Herbert Gans points out in his classic study *The Urban Villagers*,[13] and can be as unforgiving of eccentricity as the stereotypical small town is.

Already a gifted student, Suzie encountered a more cosmopolitan outlook at Urbus Orbis, whose name is, after all, "a bastardized version urbi et orbi, Latin for 'to the city and to the world.'"[14] In the late 1990s, Suzie left Chicago for the University of Pennsylvania, where she earned a degree.

Despite huge local popularity and national recognition that reached its zenith when *Rolling Stone* magazine named it "the coolest place [in Chicago] to suck down a cappuccino,"[15] Urbus Orbis was not much of a financial success, and it closed down in 1998. Urbus Orbis did match the social vision of its proprietor, as a place where a local cadre of artists and others attracted to the lifestyle affectations of bohemia interacted and developed a sense of community. In doing so, it compromised economic success. The laid-back ethos of Urbus Orbis contrasts with that of coffee kingpin Starbuck's, which serves decidedly non-bottomless cups of coffee. In 2001, Starbuck's opened the outlet long dreaded by neighborhood purists on the central intersection of North, Damen, and Milwaukee, two blocks west of where Urbus Orbis once reigned as a center of neighborhood culture.

Urbus Orbis made visible the artistic presence as well as the new potential of the neighborhood's industrial relics, potential that would be harvested by more canny business operators in subsequent years. But it also highlights the internal contradictions of neo-bohemia that arise between the need for free space in which ideas and innovations can emerge from spontaneous interactions and the relentless crush of routinization, homogenization, and rationalization that accompany the commodification of everyday life in the city.

Out of the Margins

Especially after 1993, the scene that was percolating in venues like Urbus Orbis and Phyllis' Musical Inn exploded the local boundaries and attracted a frenzy of national attention, including features in media outlets like *Billboard* and the *New York Times*. Local artist Liz Phair became the new indie rock darling, gracing the cover of *Rolling Stone*,[16] and the neighborhood that she called Guyville became a new bohemian tourist destination. It was around this time that Tim Rutilli insisted to the *New York Times* that Wicker Park was not "Paris in the 20's," perhaps hoping to dampen the romantic yearnings that might attract too many (and the wrong sort of) participants. As we have seen, such yearnings were in fact a significant part of what drew artists like Rutilli to the neighborhood in the first place, when it was depopulated and crime-ridden, providing a set of conceptual associations that allowed the gritty district to be interpreted in terms of bohemian fantasy. Such associations now have a wider appeal, however, and the 1990s brought not only a deluge of artistic aspirants, but also increasing numbers of white professionals hot to get in on the cachet of the hipster neighborhood, especially after the first wave had, perhaps unintentionally, helped to pacify its wilder elements.

Among artists and committed scene makers in Wicker Park, the attractiveness of the scene for those deemed outsiders was generally taken for granted. As Jenna Hill put it in 1995, "You know, this is a hip neighborhood. And there are a lot of people who want to come and hang out, because artists are cool. Or because art is, anyway." Typically, local artists in the 1990s took Wicker Park's cachet to confirm the value of their subcultural capital at the same time that they complained it was ruining the neighborhood. But while it is true that past bohemias, from Montmartre to Greenwich Village, drew visits from "slumming" elites, the marketability of "living like an artist" in contemporary cities is in fact novel both in terms of scale and style, and is not something that we can just assume is somehow "natural."

After all, mistrust, if not outright hostility, has traditionally characterized the middle-class disposition toward artists and intellectuals in the United States (not surprisingly for a nation steeped in cultural populism and the Protestant ethic), and the scope and appeal of bohemian enclaves were quite limited for much of the twentieth century. Given the uniformly appreciative tenor of the press' take on Wicker Park, it is easy to forget that Greenwich Village was as much a target for media condemnation as celebration during its bohemian heyday.[17] The flourishing of new bohemian United States cities, appearing simultaneously as sites of cultural creation and as distinctly themed spaces of consumption fawningly advertised by the mainstream media, thus requires explanation.

In her now classic study on the New York loft market, Sharon Zukin contends, "It is inconceivable that living 'like an artist' would have exerted any appeal to segments of the middle class if significant changes in the social positions of art and artists had not taken place since the end of World War II."[18] While New York's early-twentieth-century bohemia had been linked to radical politics, Serge Guilbaut holds that the postwar "Tenth Street School" of abstract expressionists (led by Jackson Pollock) was symbolically enlisted on behalf of capitalist hegemony in the ideological struggles of the Cold War, illustrating the virtues of freedom and creativity, even where this co-optation confounded the political dispositions of the artists themselves.[19] As Thomas Crow demonstrates, this was the period when the avant-garde first became safe for photo spreads in *Life* magazine, explicitly identified with the new extensions of mass fashions.[20] Links between art and commerce have only increased since then, helped along by avatars of pop postmodernism like Warhol, and later Basquiat, Koons, and Haring. Both the accessibility and appeal of bohemian lifestyles increase with the postmodern extension of an aesthetic economy.

The much wider appeal of bohemian signs and spaces (whether or not they are explicitly called bohemian, or attract tags such as "indie" or "alternative") confounds the tradition of anti-intellectualism in American life traced by Richard Hofstadter.[21] This anti-intellectual ethos is far from past, of course; as Andrew Ross argues, popular culture, while borrowing liberally from modern and postmodern innovations in the high arts, still generally takes a stance in the mass market that treats intellectuals with "no respect."[22] Moreover, the arts, especially insofar as they are publicly subsidized through the National Endowment for the Arts, still demonstrated ample ability to shock and alarm large segments of the American population in the 1980s, making them fine vehicles for conservative politicians waging the so-called culture wars.[23] If indeed the arts and the artist's lifestyle now appeal to a much wider United States audience, one should still not get carried away by declaring the death of the American mistrust of the cultural class.

Nonetheless, Zukin's point about the appeal of the artist's lifestyle to "segments of the middle class" is accurate and essential, though the change has as much to do with the structural constitution and composition of these segments as it does with any changes in the practices of artists themselves. The global city now concentrates a workforce that evinces high levels of education, employed to perform the kind of tasks that Robert Reich calls "symbolic analytic"[24] and that Richard Florida simply calls "creative."[25] As we have seen, the unionized, blue-collar work-force of the Fordist city, the foundation of Chicago's economy and the backbone of its democratic machine politics, has been decimated by the forces of deindustrialization and white flight. In the wake of Fordism's decline, a dual city is produced, with increased instability and immisera-tion for disadvantaged populations, largely new immigrants and African Americans, alongside increased wealth and opportunity for the profes-sional class employed in new growth sectors. Chicago's fortunes are now driven by finance, legal services, media, and new technology, sectors that employ workers whose usual socioeconomic and cultural sphere is very different from that of industrial laborers. Wicker Park in the 1990s shows the intersection of these competing trends in the postindustrial city, with disadvantaged immigrants and young urban professionals encountering one another on its cracked sidewalks.

Residential Change

New arrivals in Wicker Park, whether "starving artists" or more affluent young professionals, tend to be well educated. The absolute number of neighborhood residents possessing a bachelor's degree or higher almost tripled from 1990 to 2000. In 1990, Wicker Park, with its high percent-age of foreign-born residents, had a much higher than average (for Cook County) percentage of adults lacking a high school degree; in 2000, the percentage of adult residents with undergraduate or graduate degrees sig-nificantly exceeded Cook County norms (table 5.1).

Other indicators confirm substantial residential turnover in the decade. The 1990 census indicates that slightly more than half the neighborhood population was Hispanic or Latino (any race); by 2000, that figure had dropped to 28 percent. New residential investment was substantial. In the period between 1993 and 1998, roughly half a billion dollars in new home purchase loans were issued, with the significant majority of these going to white rather than Latino homeowners (table 5.2). Comparing census data in 1990 and 2000, we can see a staggering increase in median rents, median home values, and median household incomes (table 5.3).

Table 5.1

Educational Attainment in Wicker Park, 1990 and 2000 (Percent of Population 25 Years and Over)

| | 1990 | | 2000 | |
	Cook County	Wicker Park	Cook County	Wicker Park
Less than 9th grade	12	28	10	11
9th to 12th grade, no diploma	15	16	13	9
High school graduate	26	20	24	12
Some college, no degree	19	12	20	13
Associate degreee	5	4	5	4
Bachelor's degree	14	12	17	32
Graduate or professional degree	8	8	11	18
Total	99	100	100	99

Source: US Bureau of the Census (sixteen-tract area).

During this period, the total population of the neighborhood remained roughly constant, but the population of children decreased with the displacement of working-class families, and the employed population (people 16 and older) increased from 15,447 to 19,774. Professionals increased from 25 percent of employed neighborhood residents to 51 percent; in absolute terms, this meant an increase from 3,890 local professionals in 1990 to 10,154 in 2000. Despite the loss of children, the median age in 2000 was 31.4, compared to a Cook County median of 33.6, indicating that the neighborhood was attracting a large number of relatively young adults in professional occupations. Thus, a neighborhood newcomer in 2000 was greeted by a very different local demography than the one encountered by the many artists who broke ground in the late 1980s. Nonetheless, Wicker Park continues to be regarded as an "artist's neighborhood," both by the press and by local residents, and it is distinguished from lakefront communities like Lincoln Park for its funkier, more offbeat character.

As an Urbus Orbis regular, Borderline employee, and respected local poet, Raul is one of the few individuals to straddle the worlds of the

Table 5.2
Wicker Park Home Purchase Loans for 1993–1998 by Census Tract

Census Tract	Total Loans	% White	% Latino	Total $ Investment	Average Loan
2402	229	86	5.6	48,053,000	209,838
2403	233	66.5	26.1	43,389,000	186,218
2404	215	89.7	1.3	35,358,000	164,455
2405	462	83.3	6	69,918,000	151,337
2412	105	73	19	17,593,000	167,552
2413	143	87.4	4.1	28,538,000	199,566
2414	325	82.7	7.3	65,097,000	200,298
2415	246	80.8	10	43,722,000	177,731
2421	184	76.6	15.2	31,533,000	71,375
2422	220	76.8	18.6	38,083,000	173,104
2423	147	80.9	11.5	40,314,000	274,244
2424	57	78.9	12.2	10,972,000	192,491
	2566			472,570,000	
Average		80.8	10.9		184,166

Source: Home Mortgage Disclosure Act (HMDA), compiled by Gray Data, Inc., analyzed by UIC Voorhes Center.
Note: Excludes Wicker Park tracts 2219, 2220, 2221, 2222.

Table 5.3
Income, House Value, and Rent in Wicker Park (Dollars), 1990 and 2000

	1990		2000	
	Cook County	Wicker Park	Cook County	Wicker Park
Median household income	32,673	23,327	45,922	54,791
Median house value	102,100	109,913	157,700	400,100
Median rent	411	325	648	802

Source: US Bureau of the Census (sixteen-tract area).

Latino immigrants and the artists in the neighborhood. Raul sees clearly the complicity of the arts community in the transformation of the neighborhood, undermining existing diversity by displacing both the original residents and, over time, many of the pathbreaking artists themselves.

> Those art school kids, what they do is they move into a neighborhood, and they don't care who lives there. They don't give a fuck, and if they do, they still don't give a fuck. Because they're moving in and that's it, that's just their thing, they're going to live there, and they're going to paint there, which is a great noble idea for an artist. But in time people notice, and start buying up the buildings, and they [decide] "I'm going to rent to these artists, give them a two-year lease, then start making repairs, and before they know it I'm going to start raising the rent." And now you have major developers jumping the gun, seeing what's happening already, and investing. The taxes go up, people can't pay, people move out. The grand migration is what's happening.

Many artists express concern about the displacement of Latino residents when talking about gentrification, although as Raul suggests, these sentiments are bound up with self-interested concerns. Here, the issues are complicated, because young, culturally competent professionals can contribute to the development and maintenance of neo-bohemia, consuming its products and ambiance. The increased visibility of the neighborhood enhances the opportunities for local artists to get their work to a wider audience, and the proliferation of entertainment venues can improve employment opportunities. Nonetheless, the escalation of ground rents generates local contradiction, as it can interfere with the reproduction of a creative labor force in the neighborhood.

The necessary compromise involves young artists being pushed to surrounding neighborhoods like Logan Square, Humboldt Park, and Garfield Park — neighborhoods that resemble 1980s Wicker Park in their composition. These artists continue to avail themselves of the Wicker Park bars, coffee shops, and display venues that have a firmer hold in the neighborhood. Thus Wicker Park remains a central organizing space for new bohemian activity in Chicago even as many artists can no longer afford to live there. Artists hover in orbit, still making the scene as they drop in to work, play, and display. Escalating rents thus drive young artists outside the borders of Wicker Park; they are excluded from living in a neighborhood whose character they continue to influence disproportionately.

The new residents filling more upscale rehabs and condominium developments are the easily spotted culprits in neighborhood change and are blamed by artists for ruining "their" scene. These residents no doubt often earn the enmity of their neighbors by failing to adhere to the same norms of neighborhood diversity, instead forming homeowner associations that resist low-income housing and agitating for increased police presence. But even when the dispositions of new residents mirror those of artists, with whom they share youth and above-average educations, professionals are still lumped onto the other side of the fence in the system of cultural distinctions that remain central to bohemian self-identification.

Neo-Bohemia and the Young Urban Professional

In the ideological formulation of neo-bohemia, the yuppie is the imago of the mainstream, playing the role once held by the bourgeois shopkeeper in nineteenth-century Paris or the "organization man" of the Fordist United States. The collective construction of the yuppie, characterized by excessive simplifications, is crucial to the formation of subcultural distinction among artists and their friends in Wicker Park. As Pierre Bourdieu points out, "Nothing classifies somebody more than the way he or she classifies."[26] Though artists in and around Wicker Park are sometimes resistant to labels like "bohemian," they are much less reticent about declaring what they are not, and what they are not are yuppies. In her excellent study of the subcultural construction of the rave scene in England, Sarah Thornton makes a point that also well describes the Wicker Park scene: "Although most clubbers and ravers characterize their own crowd as mixed or difficult to classify, they are generally happy to identify a homogeneous crowd to which they don't belong. And while there are many *other* scenes, most clubbers and ravers see themselves as outside and in opposition to the 'mainstream.'"[27]

In Wicker Park, "yuppie" and "mainstream" became practically interchangeable terms. Autumn, a young actress living in the neighborhood during the 1990s, offers an exemplary local take on the figure of the yuppie:

> A yuppie would be like, in Lincoln Park or Wrigleyville, the upper middle class, materialistic, out for themselves, not very into community, not very into the arts. Into making themselves secure futures, and that's what's most important. Not that I'm saying there's anything wrong with that, but as an artist you've opened yourself up to the fact that you may not be secure for the rest of your life.

Autumn's broad view illustrates important themes in the species distinction that artists make between themselves and young professionals, including the use of spatial affiliation as a sign of membership. As Autumn indicates, artists live in Wicker Park, or a handful of other gritty neighborhoods like Pilsen, Logan Square, Ukrainian Village, or Uptown, as opposed to the gentrified lakefront communities of Lincoln Park and Wrigleyville.

But by the time that Autumn had arrived in Wicker Park, it was already fast becoming home to a large population of professionals, complicating her spatial distinctions. In some cases these newcomers would be given a pass, so to speak, by their more bohemian neighbors, so long as they performed appropriate respect for the arts. As Tom Lee puts it, "I know several people who were commodities or options or equity traders at the Board of Trade. But they belonged. One of them, even though he trades during the day, he paints through the evenings and supports his wife, who's an opera singer." For Lee, what it comes down to is "priorities — the life of the mind, creativity, is what is valued over the bank account." Few yuppies are considered to have their priorities in order to this degree, however. Most artists concur with Autumn's claim that the yuppie emphasizes security while neglecting community and creativity. This view mirrors the negative image of the organization man and Fordist ideology, even though this is complicated objectively by the changing nature of work in the postindustrial economy, in which even relatively privileged workers endure much higher levels of instability in employment.[28] Against the perception if not the reality of yuppie timidity, the bohemian ethic repudiates security and embraces contingency.

The insecurity and relative deprivation of the artists' lifestyle is often described as an advantage over the staid existence of buttoned-down professionals, and in this way artists signal the superiority of their existence over both the poor and the privileged. Says Shappy, a local performer:

> I don't think [yuppies] have any creative gumption. Yes they may take chances on a business deal or an ad campaign or something stupid . . . but they don't have the balls to put it in play in their own personal lives. And when they see people *living* I think they're jealous of the artist's lifestyle, wishing they could feel like they could be free and live on macaroni and cheese and not have to worry about these accounts and their bills and their credit cards and their SUVs, and their blah, blah, blah. You know, I think a lot of people want to be more bohemian, but they don't want to take the chance on actually living the life as a bohemian. They're too insecure without their credit cards.

Shappy finds evidence for his theory in the attraction of Wicker Park and its hipster scene for young professionals in the neighborhood, who dabble on the weekends while remaining unwilling to make his level of commitment. Still, he also acknowledges, "I live pretty modestly, and for fuck's sake, I'm 32 years old, and I'm probably still below the poverty level, and it doesn't really bother me, but yeah, it would be nice to have some chump change."

Both the materialism of yuppies and their antipathy toward community have become articles of faith, not just for artists, but also for most academics. According to the ethnographer Gerald Suttles, "The term 'yuppie' most obviously applies to young singles, who are heavily preoccupied with their nightlife, exploring the new reaches of consumerism, and staying abreast of the trends."[29] Gentrification scholar Neil Smith concurs: "Apart from age, upward mobility and an urban domicile, yuppies are supposed to be distinguished by a lifestyle of inveterate consumption."[30]

But the presence of educated professionals in Chicago has been central to the development of neo-bohemia, massively increasing the demand for cultural goods in the city, despite Autumn's stereotypical claim that they are indifferent to the arts.

The omnivorous cultural preferences of the new urban class of postindustrial professionals sit behind the development of "the city as an entertainment machine" generating a range of cultural amenities.[31] Still, while these developments have been critical to the viability of new bohemias, expanding audiences as well as employment opportunities for artists, this has not increased the popularity of young professionals. Distancing themselves from yuppies promotes in-group solidarity for the diverse cast of aspiring artists and hipster hangers-on that populated Urbus Orbis. It also allows them to downplay their own complicity in the gentrification displacing poor Latino residents. Finally, by deriding the soulless search for security, local artists make their own commitment to risk into an expression of comparative virtue.

CHAPTER 6

The Celebrity Neighborhood

> Bohemians go everywhere and know everything;
> sometimes their boots are varnished, sometimes down
> at the heel, and their knowledge and the manner of
> their going varies accordingly. You may find one of them
> one day leaning against the chimney-piece in some
> fashionable drawing room, and the next at a table in
> some dancing saloon.
> — Henri Murger, *Scenes from the Bohemian Life*, 1848

IT'S AFTER 1 A.M. on a February night in 1999, and Allan Garland and I descend the stairs to the bowels of the multi-level nightclub Stardust. We can hear the muted sounds of Chicago-style house music from the other side of its heavy doors as we are confronted by the gatekeeper, a pixyish young woman with two-tone dyed hair and a long vintage coat. Sprayed-on glitter sparkles on her cheeks in the dim light. "It's $10 [to get in]," she informs us. Allan, a slender black man in his middle twenties, rocks on his heels, dreadlocks bobbing. We had just left the downtown café Third Coast, where Allan, an aspiring club DJ and West Side resident, works. He's still dressed in his work-mandated white shirt, now untucked and mostly unbuttoned. "Does it matter if I'm *industry?*" he asks, with a conspiratorial smile.

She squints. "Where do you work?"

"I work at Third Coast," Allan explains. "And I spin." He rattles off a résumé detailing venues where he has subbed as a DJ. Her squint turns to recognition and a beaming smile. "I know you!" she says happily. "You're Norman's friend." Allan nods, and the door swings open. "Go in," she instructs, forgetting our cover charge obligation. Allan bounces through the door. I hesitate, since my own affiliation, with the University of Chicago, seems likely to have less resonance here. But she nods me along. I am to be the beneficiary of Allan's "industry" aura.

This club, a popular hangout with hip kids and affluent young professionals about a mile southeast of Wicker Park, is lodged in what Ernest Burgess called the city's "zone in transition," near the terminus of railroad tracks that long exported the city's impressive industrial output. Just past the western fringe of the Loop, the area is thick with warehouses and low-rise factory buildings. But the zone's contemporary transition involves the proliferation of living lofts, high-end restaurants, and hip nightclubs. This transition is far from total. During the day, this stretch is still thick with big trucks and prowling forklifts. But where the night once saw it mostly deserted, now its nocturnal sidewalks are lined with stylish young people and its streets patrolled by yellow cabs. And in this after-hours world, "industry" work refers to mixing drinks, serving food, or spinning records.

Sections of Chicago's West Side are evolving into a glamour zone of warehouses-turned-nightclubs, new-wave restaurants, and *noir*-themed bars. Chicago's new nightlife integrates the former industrial neighborhoods of the West Side into the global city that is also an "entertainment machine,"[1] satisfying consumer demands made by a young, relatively affluent, and well-educated workforce. As an industry, the entertainment scene has less in common with the manufacturing that once dominated the West Side, often in the same structures, than it does with Hollywood — it is a culture industry, and the contributions made by its workers are significantly, but not only, aesthetic.

In the past, serious urban scholars have minimized or ignored the city's leisure economy. But the ongoing decline of heavy industry in U.S. cities has been accompanied by an increase in the scope and impact of tourism and consumption on urban fortunes. Terry Nichols Clark indicates that "Chicago's number one industry has become entertainment, which city officials define as including tourism, conventions, hotels, restaurants, and related economic activities."[2] Urban tourism is fast becoming a central, not a tertiary, object of academic study. But by and large, new theories emphasize consumption districts that operate at a radical disjuncture from the everyday life of residents[3] — in Chicago, themed attractions like Navy Pier, or the phantasmagoric shopping spectacles along the Magnificent Mile. Dennis Judd calls such districts "tourist bubbles,"[4] and, as in Fredric Jameson's famed analysis of the Bonaventure hotel, the postmodern manifestation of culture as capital is indifferent to and alienated from the quotidian space of the city.[5]

Disneyland has become a central symbol of this new style of consumer space. Critics see Mickey Mouse as symbol of bleak postmodern dystopia — the absolute triumph of artifice over reality.[6] Argues Michael Sorkin, a leading critic of theme-park urbanism, "The empire of Disney transcends [its] physical sites; its aura is all-pervasive."[7] But whatever its postmodern features, the actual production of the Disney aesthetic adheres to the Fordist principles that Adorno and Horkheimer saw in the culture industries of midcentury,[8] with a routinized labor process and standardized output. Indeed, arguably the one place where Adorno and Horkheimer's analysis is today repeated uncritically is in the new urban criticism of Sorkin and other Disneyfication theorists. In their totalizing scope, adherents to this paradigm often seem to forget Jean-François Lyotard's admonition: "Eclecticism is the degree zero of the contemporary general situation."[9] Wicker Park challenges the "all-pervasive" saturation of these models, as flexible, post-Fordist arrangements of cultural production and labor-force exploitation characterize its entertainment economy. While Disneyland generates its aesthetic "from above," imposing rigid standards of appearance and performance on its workforce, the bars and restaurants in neobohemia piece out image construction to individual employees steeped in the local neo-bohemian subculture.

Moreover, the assumption of a homogenized urban landscape inherent in the Disneyfication thesis misses the extent to which urban tourism draws upon distinctive features of local history. In a richly textured ethnographic study, David Grazian examines how Chicago's legacy as the "home of the blues" is tapped in the active construction of the contemporary tourist economy.[10] As Grazian demonstrates, visitors to the city seek out blues clubs as a central feature of the Chicago experience, with club musicians duly offering rote performances of weathered stand-bys in order to satisfy tourists' reified expectations of authenticity. This presentation of a well-seasoned tradition does not tell the whole story, however. City residents also frequent blues clubs (often favoring those that are off-the-beaten-track), striving to distinguish themselves from tourist dabblers through their superior knowledge of the musical tradition, as well as their superior competence in nocturnal comportment.

As the residential profile of Chicago changes, city dwellers do not leave entertainment only to visitors, but use their own city "as if tourists,"[11] aggressively pursuing urban consumption opportunities. This is particularly true of the young urban professionals who fill jobs in the high-rise offices of the global economy and also drive neighborhood gentrification.[12] This "new class" is quite different in character from the blue-collar workers of Fordism; their consumption habits reflect not only their affluence but also their occupations, education, and age.[13] The ability of the global city to capture and retain "talent" hinges in part on responsiveness to their aesthetic dispositions,[14] typically "omnivorous" tastes[15] that include fondness for the Chicago Bulls as well as off-Loop theater, or at least the idea of off-Loop theater. As sophisticated city dwellers, they seek out diverse consumption opportunities beyond mass attractions like Navy Pier, Chicago's number one visitor destination.[16] For some, a neighborhood like Wicker Park brokers fantasies of a hipper, more authentic urbanism, available to discerning insiders but not the tourist hoard.

The Wicker Park Scene

As Wicker Park achieved celebrity status, promoted by sustained notice in the local media of its many hip attractions, the number of entertainment venues expanded dramatically. New bars and restaurants opened regularly in the neighborhood throughout the decade and beyond, although, as is often the case with small businesses, many of these survived only a short while. The staples of the emerging hipster scene were the Rainbo Club, Bop Shop, Phyllis' Musical Inn, Borderline, Czar Bar, Innertown

Table 6.1

Concentration of Selected Industries in Bucktown and Wicker Park Relative to the Chicago MSA, 2000

Industry Code	Industry	Location Quotient
31–	Manufacturing	1.3
711110	Theater companies & dinner theaters	3.1
711130	Musical groups & artists	2.0
711510	Independent artists, writers & performers	1.7
722110/211	Full and limited-service restaurants	1.6
722410	Drinking places (alcoholic beverages)	1.8

Source: U.S. County Business Patterns, 2000.

The Location Quotient (LQ) statistic can be used to measure how concentrated an industry is in a particular geographic area at a point in time. An LQ greater than one indicates that the concentration exceeds the average for the Chicago Metropolitan Standard Area. Thus, Drinking places has an LQ of 1.8, indicating that the share of drinking places in Wicker Park is eighty percent greater than in the MSA.

Pub, Artful Dodger, Sweet Alice's, Hothouse, North Side, Dreamerz, Subterranean, Red Dog, and Gold Star. Later arrivals in the neighborhood area include Lava Lounge, 1056, Double Door, Holiday, Nicks, Mad Bar, Bigwig, Davenport's, Eddy Clearwater's, Pontiac, and the Note. Thus, by 2000 the neighborhood, while not completely bereft of its older industrial character, had become a hub of culture and entertainment, as an examination of the location quotients for relevant industries reveals (table 6.1).

The patronage of these bars is hardly limited to committed participants in the neighborhood arts scene; for the most part they are populated today by college students and young professionals — individuals with enough disposable income to actually make these businesses viable. Nonetheless, the hip, neo-bohemian ethos of the neighborhood remains thematized in this nightlife scene. Several of these bars feature live music, and others host open-mikes for local poets and writers to read their work. Walls are often decorated by murals or hanging pieces created by local visual artists. Interiors are typically kept deliberately unpolished and are decorated with retro furnishings like old couches and lamps. Usually, they are dimly lit, with lighting strategically deployed to produce the shadowy, chiaroscuro effects associated with film noir.[17] Moreover, the connection to the arts community in these bars is displayed through the personae of the bartenders and wait staff, which disproportionately consists of young cultural producers and aspirants.

In addition to these watering holes, the neighborhood is now also thick on the ground with restaurants that often cater to an affluent as well as stylish clientele. These venues are representative of the various types of "new wave" restaurants that Zukin notes began to multiply in big cities during the 1980s,[18] reflecting the cosmopolitan tastes of professionals in a globalizing economy. "Restaurants have become incubators of innovation in urban culture. . . . For cultural consumers, restaurants produce an increasingly global product tailored to local tastes."[19] Spring, a popular and high-priced restaurant on North Avenue, serves Pan-Asian cuisine, a style that quotes ethnic culinary traditions rather than reproducing them. For several years in Wicker Park only the decidedly mediocre Pacific Café served sushi, a pricey bill of fare that is extremely popular with young professionals. By 2001, there were five sushi restaurants in the neighborhood: Blue Fin, Mirai, Bob San, Papajin, and the inexplicably still-open Pacific Café. Moreover, a range of other regional and ethnic cuisines are readily available: Thai (Thai Lagoon), Persian (Souk), Italian (Babaluci), Soul Food (Soul Kitchen), and many more. Lodged in converted industrial spaces and retail storefronts, these restaurants display the dramatic juxtaposition of grit and glamour that is a key modality of neo-bohemian value. And despite the cosmopolitan cultural influences, the restaurants' public face is presented by servers who are American-born artists, marrying exotic cuisines to the hip and funky ambiance of the local bohemia. Thus, many of these restaurants thematize, all at once, culinary tastes emblematic of postmodern globalization (simulacra of "authentic" ethnic traditions), the adaptive recycling of postindustrial symbols and spaces, and the creative vibrancy of an artists' community.

Both bars and restaurants cluster along the neighborhood's main drags of Milwaukee, North, Damen, and Division, interspersed with new retail outlets, as well as the check-cashing stands and discount furniture stores that endure from a less high-profile period. They are key pieces of

Figure 6.1 Sidewalk Seating on Division Street. Photo by the author.

the "retail renaissance" that includes shops selling antiques, art supplies, records and CDs, and fashionable (even edgy) attire, along with some two dozen art galleries.[20] Actual artists, as well as young professionals, live above these establishments in Milwaukee Avenue lofts, for example. These mixed uses produce just the kind of lively and eclectic pedestrian traffic that Jane Jacobs lauded with so much enthusiasm in her description of Greenwich Village;[21] they belie dystopic imaginations of the dead inner city. Thus is the "scene" constituted in neo-bohemia, through the activities of artists, entrepreneurs, consumers, and service laborers, categories of social actors that bleed into one another.

The popularity of the neo-bohemian scene, taken as whole, demonstrates its ongoing appeal to a new class of urban consumers even though, or perhaps because, the gritty motifs discussed in chapter 3 increasingly belie the reality of a more upscale residential and consumption profile. For all the demographic changes of the past ten years on Chicago's near West Side, there persists the allure of the cutting edge on which local entrepreneurs capitalize, making use of local artists as standard-bearers in the process.

Hiring artists keeps the businesses tapped into an ethos of hip creativity. Moreover, entrepreneurs in the neighborhood justifiably view themselves as creative scene makers in their own right. By opening and maintaining establishments that express the neo-bohemian place ethos of Wicker Park, they do not merely appropriate elements of the scene — they are integral components of the scene.

From the North Side to Mirai

By examining the cases of two establishments, opened at opposite ends of the 1990s, we can see the continuity in strategies of scene production in the neighborhood. When Cyril Landise opened the North Side Café in 1988, it was among the first new businesses in Wicker Park clearly banking on neighborhood change. In contrast to the dense entertainment scene depicted above, Landise recalls, "All the storefronts from North Avenue to the tracks were vacant, every single storefront." Although the bar, with its fake fireplace and beer garden, resembles many such venues catering to comparatively affluent professionals in the city's lakefront communities like Lincoln Park, Landise insists that the relative underdevelopment of the neighborhood allowed him to go for a funkier vibe, consistent with the emerging neo-bohemia, and predictive of subsequent neighborhood designs. When I interviewed him in 1995, he recalled:

> The advantages were that it was a little more free form. Since the neighborhood wasn't sedate . . . there weren't a lot of restrictions on what we could do. So it just felt freer, we could make a little more noise, we could have a little more bizarre [attractions]. We could book bands once in a while, and they would play out in the garden. We'd try different kinds of music, we'd try more bizarre menu items. We were able to do a lot of things you didn't do at Bennigan's because it wasn't done. We didn't care if it wasn't done, in fact we liked it better if it had never been done.

Like Urbus Orbis' owner, Tom Handley, Landise liked to think of his bar as far more than a business opportunity. Though he did not explicitly

Figure 6.2 Mirai. Photo by the author.

mention Oldenberg's book *The Great Good Place*, Landise may as well have been quoting from it when describing his aspirations for the North Side. Oldenberg argues that third places serve as social levelers, and Landise asserts, "The most distinctive expression that I try to put in every fiber of the place is a sense of equality, that is the fact that I want this to be a respite from any sense of the caste system. . . . My idea of a really interesting place to be is somebody with purple spiked hair sitting next to a guy in a suit talking about some topic of interest at the bar."

For Landise, as for many other subsequent local entrepreneurs, the growing population of neighborhood artists made available a workforce consistent with his goal of a more "free-form" environment:

> Everyone who worked here really saw it as a mixture of lifestyle and income. Anybody who sees it as just income doesn't last long. They tend to come in — the corporate waitress types, the bartenders, the managers who are looking for a gig — [but] for some reason don't stay here. The ones who stay here are people who are artists and

actresses, writers. There's almost a sense of community here. And they get to talk with people, there's much less distinction between a patron and a server here.

Landise claims that the high number of artists employed by his estab-lishment did not occur by conscious design. He concedes that "people who are artists are more interesting and intelligent than those who aren't" — presumably an advantage in the interview process. But he adds, "I never consciously sought out artists. They'd come in here and they'd apply, [but] we never said 'Are you an artist?' We'd hire them for totally different reasons and then find out that they are." Landise claims to hire both artists and non-artists, while conceding that artists are easier to retain since they are more likely to buy into the employment culture that he seeks to maintain. In doing so they become part of the scene: not just passive servers catering to the whims of patrons, but active participants in the production of the overall ambiance, the creation of a "really interesting place."

Granted these workers differ from the other participants in the scene by virtue of having to take orders, serve food and beverages, and mop counters in addition to engaging in spirited, third place–style exchanges. But Landise is right to note that they contribute more than base servility. Being "more interesting and intelligent than other people," and almost always possessing exceptional competence at the key leisure pursuit of fashionability, the artist as service worker contributes to the production of the bar's ambiance in an outsize way. This double duty comes at a cost to Landise of minimum wage, which, as he proudly points out, is twice what he is legally mandated to pay employees who work primarily for tips. The vaunted ideology of "community" and "creativity" becomes the coin with which this labor is secured. Landise tells the story of an employee who came to the North Side from a much swankier and more remunerative restaurant job:

The waitress right there in fact came here from Charlie Trotter's, a very high-end restaurant on Armitage, $250 for two people to eat. It's one of the few four-star restaurants in Chicago, and she was a waitress there, and she said, "No, that's not for me. That's a lifestyle commitment to servitude that I'm not interested in subscribing to, even though the money's great." So she can come here and make less money, but not be a server: she can be a co-equal in the community.

And yet, service workers at the North Side and other neighborhood bars are not co-equals with either Landise or the bar's patrons — within the structure of the social situation, they remain subordinates, although the way that subordinate status plays out can be complicated.

This is not to presume that Landise is being insincere, and as Erik Olin Wright points out, small capitalists occupy structurally contradictory positions with regard to relations of production.[22] Unlike in large corporations, the owners typically toil in close proximity to their employees, often realizing small and precarious financial rewards for their labor. While the corporate form allows top executives to insulate themselves from risk that is passed on to shareholders and employees, this is not case for entrepreneurs like Landise. His personal stake in the venture far exceeds that of his employees, who do not view their employment as a permanent commitment, and who typically move from one service job to the next with some regularity before departing the service "industry" altogether. While Landise may yearn for community, in order to be viable in a competitive arena, his business must eventually extract significant surplus value from the labor of its employees, and therefore must direct its strategies toward that end. As it happens, this is a goal many small businesses fail to achieve, leading to short average life spans for new enterprises. Landise points out that for the first two years of the North Side's existence it was not profitable, a fact that created enormous personal strain:

> I got a developer to essentially lend me a couple of hundred thousand dollars on a handshake, and that worked, and within a couple of years we were a viable business. In two years I was actually paying myself regularly. For about a year and a half I didn't get paid so often. The staff always got paid. The owners didn't always get paid. So I lived off credit cards. I was very much in debt. . . . So you learn to lay awake at night and watch the trees blow, and say, "My God."

His employees have much less invested in the success of the venture; they view their employment as temporary, and therefore, while most would wish him the best of luck, they have far less personal commitment to the business. Genuine desire to produce a harmonious workplace intersects with the requirement that Landise control his workers, however soft and apparently benign the strategies directed to this end may appear. And the hard fact remains that Landise has the privilege of being able to terminate the employment relationship, a decisive source of power asymmetry. At the North Side, as with other bars and restaurants in the neighborhood, rates of employee turnover are quite high, despite the ethos of community and collective enterprise.

The North Side persists in its Damen Avenue location, now surrounded by a host of trendy competitors, and it continues to employ a rotating cast of colorful employees linked to the arts in Chicago. For all the changes in the local retail ecology, new businesses continue to echo the strategies articulated by Landise, actively striving to construct a sense of cutting-edge ambiance. Mirai, a sushi restaurant, opened in 1999 at the opposite end of the neighborhood, on what was then a lightly developed stretch of Division Street just west of Damen. Like the North Side, it helped to initiate what has become a fairly dramatic retail expansion in its immediate vicinity. When Mirai opened, the development on the street was still decidedly uneven, with a Laundromat across the street and vacant lots and shuttered storefronts interspersed on the block.

Mirai is a chic, two-story venue catering to the beautiful people, as Matt Gans, its initial manager, indicates:

Mirai is crazy, man. It plays to this obsession with the Asian culture. They have the finest sushi chefs in the country. They put this funky nightclub upstairs and the whole thing was designed by this crazy French guy Francois, so it was ultra-swanky. But it's a restaurant, not a nightclub; people go upstairs to wait [for a table]. They don't mind [waiting] because the upstairs bar is this totally swank fancy hangout, [people in] great clothing. They come dressed to kill. It's a stomping ground, it's a place they want to come to be seen at before they go out to be seen. . . . It's an insanely good location. This area had not had that caliber of restaurant, as long as everybody's been living here. They had Pacific Café, which is also sushi, [but] awful sushi. So they'd been kind of waiting for something to come to the neighborhood. In a way, it's completely different for Wicker Park, it's not grungy — it's not anything you would expect around here.

But the juxtaposition of Mirai's posh atmosphere with the underdeveloped local landscape serves to heighten the drama for its patrons, and allows them to imagine that they are consuming a product unavailable to those too timid or uninformed to venture into the wilds of the new bohemia.

Geared to hip and well-heeled party people, rather than greasy slimy artists, the link to the neighborhood neo-bohemia is manifested in the staff. Says Gans:

[The staff] is very representative of Wicker Park. The girls are very artistically inclined, very imaginative, very creative. The guys are the same, they're musicians. We specifically hired very funky-looking people because of wanting to appeal to Wicker Park. Meia [the owner] specifically hired these girls that looked crazy. They looked totally different than everybody else. It's eye candy. It's something you don't normally get up close and personal. Piercings, different colored hair, no bras. *It's what's going on.*

This hiring strategy was thus directed explicitly to producing the desired ambiance at Mirai — that is, to making the scene. As with Landise at the North Side, the hiring of artists is also a move toward securing employee loyalty and engendering the kind of soft control associated with the post-Fordist workplace. Gans elaborates: "It was strategic for us to hire from this neighborhood, people who live here that can walk to work. We thought they would be very prone to give us their souls because they lived here." As for their artistic interests, he says, "I think it makes a difference, because a lot of these people are so passionate about what they do, as far as their creativity and what they do personally, and [they like] being able to come to a place where they can kind of express that, where they feel they can be who they are."

Neighborhood resident and musician Brent Puls was an early employee of Mirai. A vocalist and saxophone player in the local funk and hip-hop band Bumpus, Puls is a handsome, hip-looking man in his mid-twenties with spiked blonde hair. Previously he had bartended at the Note on Milwaukee Avenue. Puls confirms Gans's observations concerning Mirai's hiring strategy:

> [The owner] hired people more for their looks than their experience. Like I'm pretty sure one girl she hired because she had dreads and was a little bit weird. It's very eclectic. We have the most handsome guy I've ever met. And then there's Larissa who has pink hair. And everyone is doing art of some kind. . . . Look at the restaurant itself. It's very sleek, somewhat elegant, but not overly fancy. And sort of contrast all the [employees] dressed in black, and there's something weird about them. The restaurant is called Mirai because it means future in Japanese. Meia wanted to be really cutting edge, the cutting-edge place to get sushi in the city, to have this really elegant place that also had these hip kids working at it.

Unlike most neighborhood venues, Mirai features an employee dress code, albeit a minimalist one. Employees were required to sport a strict ensemble consisting of a solid white shirt and solid black pants or skirt, items they were expected to supply themselves. This stark ensemble serves to highlight the striking features of the individual employees, from stunning good looks to dreadlocks or pink hair.

Gans concedes that Meia, the owner, evinces a highly driven and authoritarian style. He saw his managerial role as requiring that he soften Meia's demands for efficient performance in communicating them to the employees, since, being artists, they are often sensitive sorts. He is an extremely affable and likable individual, and he presents a laid-back persona even when giving strict instructions. Still, he also points out that his effort to cajole employee's hard work in a friendly fashion was backed up by the power to fire them if they failed to go along:

> Meia, she's the Gestapo owner, the Fuhrer, the drive, y'know; she's relentless. But she also has the respect of everyone who works there

because she's a woman in this business by herself that's made a name for herself, and they're making good money. I took everything she said and put it in terms that they could understand. And I always had a smile on my face. But I was also the person who fired every single person [who got fired] in that place. I was your best friend, and I would help you out as best I could, but if you weren't doing your job, I was also the person to fire you. I would cut you off right away.

Precisely because Wicker Park and the surrounding neighborhoods remain beacons for young aspirants in the arts into the 2000s, there is more than enough slack in the labor market to swiftly replace those who are "cut off" with similarly colorful new workers.

Scene Makers

The neighborhood demography and the elective affinity between the flexible lifestyle dispositions of young artists and service work connect entrepreneurs to a steady stream of applicants from the arts community. Established bars and restaurants rarely take out ads for new employees. Information about new openings travels by word of mouth, and preference is given to the acquaintances of current staff in further hiring, buttressing the sense of community. Hiring quickly becomes a closed loop, dominated by artists and their friends, a state of affairs that generally suits the interests of employers nicely.

Thus, new employees typically enter into these jobs by leveraging contacts that they have made in the arts community. For example, Amy Teri, who worked first as a cocktail waitress and then as a bartender for almost a decade at the Borderline, was invited to apply by a classmate at Columbia College. "I got the job when I was 21. Do you remember Amy Novack? The staple of the Borderline. Actually I had a photography class with her. They needed a cocktail waitress, and I started working as a cocktail waitress." Tall and extremely striking, Amy Novack worked at the Borderline throughout the 1990s, leaving a gaping hole in the local scene when she eventually departed for New York toward the end of the decade. Together, the two Amys (known by regulars as "Big Amy" and "Little Amy") were fixtures behind the bar, with their outsize personalities a key part of the ambiance. By hiring their friends, they stamped the place with an enduring personality. Amy Teri eventually took charge of the hiring herself, and she adopted a strategy similar to Amy Novack's when it came to recruitment: "It's pretty much, people that get hired here are regular customers or they know someone, it's a friend-of-a-friend kind of thing. We never, ever hired anyone who has walked in off the street and said, 'Here's my application.' I just throw them in the garbage. . . . Everybody who works here now knew somebody, and that's how they have the job here."

This pattern not only keeps the Borderline's owners (three Macedonian brothers who also own Café Absinthe, Red Dog, and the Blue Dolphin) tapped into a labor market comprising cool kids with high neighborhood profiles, but it also buttresses the kind of "soft control" that prevails in the post-Fordist workplace. Amy Teri adds, "It's nice, all the bartenders are pretty close and hang out after work, and it's a pretty tight family here. That's about all we have going on right now. If you don't have that behind the bar, then forget it, then otherwise it will drive you crazy."

These employees also improve the ambiance by serving as magnets for their friends, who likely possess hip cultural capital, and by frequenting the bar on their off nights. Young artists and service workers like to visit establishments where they know members of the staff, as the special attention they receive is one way insider status in the scene is confirmed. Many bartenders in the neighborhood are a source of attraction by virtue of their exalted position in the service status hierarchy. Local writer and bartender Krystal Ashe describes them as "the Startenders," a common phrase in the industry. "There are bartenders that work at three or four different clubs, [and] that promote their own nights, and people go there because they are working there." This dimension can be crucial to the bar's ambiance, since patrons measure one another in determining the coolness of the bar or restaurant.

Dealing with the Amateurs

Most bars and many restaurants are only open during the evening; in any event, nighttime is usually where the bulk of activity occurs and the most money is made, for owners and staff alike. If they are available at all, day shifts are comparatively unprofitable and undesirable. They can also be depressing. Krystal remembers, "I had a day shift [in a neighborhood bar], and people would be coming in five days a week. That lasted three weeks for me because it's just so sad." The Borderline opens at 2 o'clock in the afternoon, and each member of the permanent staff must work one afternoon shift a week. Most of the work on this shift consists of stocking and preparing the bar for the evening, a duty carefully attended to out of a sense of obligation to the co-workers who will replace them on the evening shift. What patronage there is generally consists of older men from the neighborhood drinking cheap beers, often regular customers who bring the bartender face to face with the unhappy consequences of a lifetime of alcoholism. These shifts are endured grudgingly as a necessary sacrifice for the right to work the far more lucrative evening hours, especially on Friday and Saturday.

Though night shifts are where the money is, they are not without drawbacks. Evening patrons do not typically produce the sadness inspired by day-shift alcoholics, but they do animate a host of status antagonisms. The night brings servers into contact with urban dwellers who are not involved directly in the arts and who hold 9-to-5 jobs, a group often lumped together under the pejorative of "amateurs" by the savvy industry professionals. This is especially true as the neighborhood demographics shift, and the vaunted entertainment scene receives constant advertisement to the metro area in the daily newspapers. Patrons who are presumed to hold professional jobs and live outside the neighborhood are especially reviled for their multiple inadequacies in the performance of bar etiquette.

The bohemian dispositions of workers can lead to a reversal of the ordinary patterns of age, race, and gender discrimination, as Landise indicates:

It's funny, in a lot of venues you have to remind the staff that they have to treat someone who's dressed in an unusual way with respect and dignity. Here you have to give the other admonition. Just because they have suits on doesn't mean they're jerks. You have to treat them politely, and if they behave properly — I don't care who they are, how old they are, what color they are, I don't care if they're

white fat businessmen in suits — you've got to be nice to them. . . .
And it's a kind of atypical admonition, usually it's the other way
around. With the staff we have attracted, the tendency is not to
want to wait on a woman in a fur or a guy in a suit. They don't treat
them very nicely.

While a clientele of professionals improves the profitability of the
establishment, these "outsiders" displease bar and restaurant workers by
failing to tip in an extravagant manner, despite the fact that they presumably command fabulous wealth. Shortly before she quit and moved to
New York, I asked Amy Novack how business was at the Borderline. She
rolled her eyes: "These fucking yuppies that come in now, they don't
know how to tip."

These "yuppies" also offend bar professionals by failing to "handle
their liquor," often becoming "sloppy drunks," especially on Friday and
Saturday nights, the big nights out for those who hold 9-to-5 weekday
jobs.[23] In addition to incompetent drinking, they are considered to operate
at a significant style deficit compared to the superhip kids that fill service
positions. Says Puls of the Note:

> You start getting the suburbanite yuppie crowd. Tight jeans. The big
> over-polished white people. The Note was a big meat market, and it
> became even more of a big market with Gold Coasters and actually
> a lot of suburbanites coming in. Arco, the bouncer, at the end of
> the night he would start kicking people out, and he would yell,
> "The bus to Schaumberg [a North Shore suburb] is leaving now!"

The Borderline continues to draw a very diverse crowd, one that is
inflected by the presence of the Red Dog nightclub upstairs. Because
Red Dog remains a hip spot on the club kid circuit and showcases different
musical styles on different nights, it draws patronage to the corner from the
black and Latino communities and from the gay community. The latter
especially come in on Monday nights, which Raul refers to as "a supreme
queen scene" popular with transsexuals and drag queens. Still, Borderline
rests at the heart of the West Side entertainment scene on the six corners,

and its patronage has also shifted from mostly artists and scene makers to increasing numbers of young professionals. While many of these may actually now live in Wicker Park or nearby, given the levels of gentrification, they are nevertheless ordinarily presumed by the staff to come from even more gentrified neighborhoods like Lincoln Park or from the suburbs, as Anne, a former cocktail waitress, indicates:

> I guess you get a lot of people from the suburbs. There was a pub-crawl the other night. Are you fucking kidding me? These guys come in: "Yeah, I want five pitchers, can we get a discount on this? Do you have plastic cups?" "No. Drink your beer and go." I hate to be a bitch, but it's like, you get these people. This girl comes in Friday night, right when I got to work, and she was drunk already. She was like, "I lost my wallet." She thinks it fell under the table. I don't know what I can do. . . . They want me to get down on my hands and knees to get their fucking wallet.

The claim that these outsiders demand inordinate, and demeaning, servility recurred often in my conversations with Wicker Park's service workers.

Nina Norris once managed the Note and later bartended a couple of blocks down on Milwaukee at the Holiday Club. She described the differences between "yuppie" patrons in her bar and cooler kids "from the neighborhood": "There's a certain kind of dress thing. The people from Lincoln Park have this weird kind of college thing. They all wear their little uniform with their sweatshirt and jeans, or their oxford and jeans, it's really kind of odd. . . . The people that live in [Wicker Park] tend to be funkier and trendier; they have their goatees, nail polish on the men, earrings and tattoos, that stuff." These descriptions show us once again that presumptive spatial affiliation becomes shorthand denoting normative distinctions around style and demeanor: people from Lincoln Park are conformist and unhip, and people from the suburbs are even worse. Of course, the 2000 census shows that half of Wicker Park's employed population was made up of young professionals. It is unlikely that more than a small number of the customers ever actually disclosed to Anne or Nina where they lived, but that doesn't matter, since "people from Lincoln Park" or "people from the suburbs" is less a geographic than a species distinction. Many of the Wicker Park artists and food and beverage service workers were themselves born and raised the suburbs, but they have repudiated this pedigree to live *la vie bohème*.

In contrast to Milwaukee Avenue bars like the Holiday Club and the Note, Mirai tends to get a more stylish crowd, although their style is not necessarily of the variety that Wicker Park artists personally identify with. Recalls Brent: "It's definitely a Gold Coast crowd. It's gonna be trendy club kids, everybody shops at Club Monacco, everyone probably on weekends does [the drug] Ecstasy at the clubs, and is probably twenty-five to thirty-five and has a lot of money. That's pretty much the clientele in a nutshell. At first it was entertaining, now it all kinds of grates on me. It's not my crowd. I can't relate to them at all." With the Gold Coast, we get another geographic designation that stands in for social type. "Gold Coast" patrons, whether or not they actually have an address in the ultra-expensive lakefront district, are trendier and flashier; they also are "better behaved." That is, they're comfortable with and well versed in norms of comportment in places such as Mirai's. Still, they are resented for their presumptive entitlement compared to the plight of "starving artists" toiling to serve them.

Nonetheless, they are better than those typed in the "suburban" and the "Lincoln Park" phyla, which err in multiple ways. Brent indicates that as well as lacking style, many patrons at the Note were clueless when it came to reading the cues of social interaction in the bar, often because of their incompetent handling of alcohol: "I saw so many things that depressed me when I was working. You know, some girl just sitting there by herself, and some guy walks over and starts talking to her, and they're both like sloppy drunk, and she's not talking to him, and he's not getting the hint."

Krystal worked at the now-defunct Mad Bar in Wicker Park when I interviewed her in 2000, and she indicated the crowd was far better behaved there, a fact that she attributed to the patronage of artists and service workers — categories that overlap:

KA: The people that come in [Mad Bar], even if they drink a lot, I hate to say they're probably alcoholics which is why they're so well controlled, it's definitely a nicer crowd. A better tipping crowd. I had to do a lot less work to make the same amount of money.

RDL: How do you account for that?

KA: More service industry, being in the middle of an artists' neighborhood, which is service industry. Here, you have people, even if they're in college, they are probably working at a restaurant to make ends meet, whereas in Lincoln Park, those people were being funded by their parents. There's a whole different respect there.

Krystal's emphasis on "respect" alerts us to patterns of behavior among some patrons that prove far more threatening to the bar workers' sense of selves than simple bad drinking and bad outfits. It is widely agreed among service workers in and around Wicker Park that "yuppies" are more likely to force workers to confront the servility that is, after all, the nature of their jobs. Says Rainbo bartender Jimmy Garbe of the "new class" of patrons at the Rainbo that appeared with increased gentrification and neighborhood notoriety, "Maybe I'm a little paranoid, but there's a definite sense that the newer clientele looks down on us a little bit. I don't want to be looked down on like I'm a servant to the person because they're dropping cash in here."

Wicker Park Gets Real

No new residents garnered as much attention — or overt animosity — as a small group of inordinately attractive men and women in their early twenties who moved into a loft conversion in the old Urbus Orbis location during the late summer of 2001. They were the cast of the cable network MTV's reality television program "The Real World." With the arrival MTV's cameras, the entertainment potential of the neighborhood exploded the local boundaries, the Wicker Park scene now figuring prominently in an immensely popular series that doubles as an instructional manual in consumer culture for its young fans.

Each season of the program takes six new young people, selected around MTV's distinctive criterion of diversity — two nonwhite (one black, one mixed race) people, one gay man, one bisexual woman, one small-town girl, one Ivy League frat boy, all great looking, all younger than twenty-four — and makes them roommates in a tricked-out apartment in the "cool" part of a major city.[24] The roommates are initially unknown to one another, and an implicit requirement for inclusion seems to be the absence of normal coping skills. At least one of the heterosexual cast members will be homophobic (in the Chicago case it was more than one), and at least one of the white cast members will be uncomfortable with black people. These attributes are aimed at generating the intrahouse turmoil that provides the dramatic fuel of the show.[25]

"The Real World" filmed its first season in New York in 1992, and has since been staged in such cities as Seattle, San Francisco, and New Orleans. The show has been an unqualified success for MTV and its parent corporation, Viacom. It has drawn increasingly higher ratings throughout the decade of its existence. According to the *New York Times*, "Its 2000 season was the highest rated show in its Tuesday night time slot on basic cable, and drew ratings among 12–34 year olds that were 173 percent higher than in its first season."[26] The "reality" format it pioneered has spawned a host of imitators, including such major network hits as *Survivor* and *The Bachelor*. These entries have varied premises, some combining the conventions of the game show while ratcheting up the stakes with around-the-clock access to contestants' lives; what they share is the use of nominally nonprofessional performers as their on-screen "talent." This means that reality cast members are unprotected by the major entertainment unions (the Screen Actors Guild and the American Federation of Television and Radio Artists) and are extremely disadvantaged when negotiating salaries.

Like "The Real World," other reality programs uniformly perform best among viewers under thirty, much to the delight of advertisers. They earn huge profits for their creators, profits inflated by the fact that they are extraordinarily inexpensive to make. Writes Bill Carter in the *New York Times*, "The fact that reality shows were far cheaper to produce than scripted sitcoms and dramas was a distinct part of their charm for network programmers. For example, a half-hour episode of 'Spy TV' costs NBC about $400,000 an episode to produce, compared to about $5.3 million an episode for the [top-rated situation comedy] 'Friends.'"[27]

This difference is accounted for primarily by the dramatic disparity in compensation for the performers in these respective shows. Lee Bailey reports that one unnamed cast member of a past "The Real World" disclosed that he received a onetime $5,000 fee for his months under scrutiny, with no further proceeds from merchandising or from syndication. As MTV prepared to sell "The Real World's" syndication rights in 2001 — a source of pure profit from which past cast members are excluded — approximately half of "The Real World" alums participated in filing a grievance against the company. But as one network executive said candidly (though anonymously), "These people are not represented by any unions — they have very little power and there are literally thousands of people behind them waiting for their fifteen minutes. I would be shocked if there was an attorney anywhere willing to take a case like that."[28]

According to *New York Times* interviews with other people affiliated with the show, the executive was indeed correct about the size of the reserve army of aspirants in line behind those lucky few selected to earn four figure incomes for appearing in a season-long television program: "'Everybody wants to be on 'The Real World,' said Maggie Malina, MTV's vice president for original TV movies. More than 35,000 people send in audition tapes each year. 'They get so desperate that they leave notes on our cars, bring their art projects to our offices,' said ["Real World" co-creator Jonathen] Murray."[29] The allure of having one's own life turned into an entertainment spectacle appears to be quite compelling in the media- and culture-saturated contemporary landscape, enough of its own reward to justify whatever self-sacrifice may be involved in ceding weeks or months of one's privacy. A further motivation appears to be found in the nebulous hope that MTV exposure will result in future culture industry rewards — a disproportionate number of cast members are aspirants in the arts, neo-bohemians. A decade of evidence shows that these hopes are largely unfounded. Generally, "Real World" cast members develop huge cult followings that dissipate as soon as the next round of participants takes the stage.

This provides a fine example of the post-Fordist strategy in which work gets renamed as something else and thus can be paid a discount wage.[30] An actor performing in front of a camera is doing work; reality performers are simply living their "real," if not their ordinary, lives. Thus, reality producers can avoid not only the compensation rules negotiated by, say, the Screen Actors Guild, but also those stipulated in U.S. minimum-wage laws. Since participating in "The Real World" is about fun or creativity or self-promotion — anything but earning money — and since cast members are so obviously eager to participate in their own exploitation, there is no problem. Viacom, on the other hand, is in it for the money. The genius of MTV was to recognize the size of the available pool of attractive, camera-savvy individuals willing to make this bargain.

The cast members are not confined to their swanky apartment, and the camera follows them as they go to work, hit the observation deck of the Sears Tower, or pursue their feverish night lives. In fact, producers indicated that the Wicker Park location, like past locales in other cities, was selected because of its proximity to a hip local bar scene. Thus, the distinct neo-bohemian aesthetic of Wicker Park, produced through the activities of countless participants not on MTV's payroll, becomes a source of value for the program. Though this was in some sense a role for which the neighborhood had been auditioning for more than a decade, the uninvited intrusion by MTV was met with something short of enthusiasm by many who remained committed to the dream of the neighborhood's neo-bohemian autonomy from corporate culture. As with the older protests against the ATC, the rhetoric invoked to disparage MTV was anti-gentrification; still it is not hard to detect a deeper resentment at the co-optation of a scene over which many hip young people feel proprietary.

Rumors circulated for months among Wicker Park artists about MTV's impending arrival, generating the familiar emotional antinomies of self-satisfaction and outrage. When word surfaced of the location, the relative old-timers who could remember Urbus Orbis viewed the choice as virtually sacrilegious. In the week before filming was scheduled to begin, a gang-style execution took place in the Burger King parking lot on Milwaukee Avenue, a few blocks from "The Real World" loft. This grisly development was greeted with morbid pleasure by neighborhood stalwarts, many of whom echoed Chicago police officer Frank Cappitelli's sentiment: "They ["The Real World" producers] got a little more real than they expected."[31]

While the majority of local residents seemed content simply to bad-mouth MTV, something they had been practicing for years, a few went further, actually staging a protest/prank aimed at disrupting the show during the first weekend of filming. Greg Gillam, a Chicago poet and writer, posted an extended account of this surreal event on his literary website, Fengi.com, edited and excerpted here:

Saturday, July 14, 2001

Around 11:30 p.m. I left an absurd, amazing show at the Empty Bottle to see another absurd, amazing thing — a crowd of 200 plus hassle the newly arrived cast of *The Real World*. About a week ago *The Real World* cast and crew arrived in Wicker Park, which displeased some current and former residents. So an underground cabal of artists pulled off a little prank. On Saturday someone distributed fake cards from MTV, inviting everyone to a party and casting of extras for the show at 11 p.m. By 11:30 between 200 and 300 people filled the sidewalks and street around the building.

Half the gathering was curious or clueless, and they provided cover for the activist half. The crowd blocked traffic, and more people got into the spirit of fucking with *The Real World*. Drums were pounded, slogans shouted, and cast members were jeered as they came and left. At one point a huge number were chanting "We're real, you're not." Cans and bottles were thrown at the building. One cast member was pelted with paper, and a red paint bomb splattered on the entrance.

The protest seemed to surprise the producers. In a way, the stupidity of their location choice is amazing. In Chicago, Wicker Park is known as the hangout of surly art types, even if rents have pushed them south and west. The Autonomous Zone, the longest lived anarchist/punk collective in Chicago, started in Wicker Park. In addition, North Avenue is central to the area's business district. The building is less than a block from North/Milwaukee/Damen, the busiest intersection in the neighborhood. It was almost obscenely conspicuous. Plus, "The Real World" building has a symbolic past. It was home to Urbus Orbis, a legendary coffeehouse and theater — a vast friendly space, a crossroads for art communities around the city. After helping to popularize the hood it was one of the first places to fall. It turned a profit, but the landlord chased it out, dreaming of wealthier leases like "The Real World."

Before I bicycled back to the more joyful Empty Bottle, I saw a perfect moment. One cast member returned to the building. He stared at the door splattered with red and then hung his head, acting genuinely upset. Then, as the crowd jeered, he held the pose so the camera could get his "reaction" from several angles. He went inside, we played our role as the angry mob, and I left. I got to the Bottle in time to see a kosher rap/funk band play Policy of Truth.

Local businesses demonstrated mixed reactions to "The Real World." The venerable used-book store, Myopic, refused to sign a release form allowing MTV to shoot on its premises and posted a sign in the bookstore's window proclaiming a "Reality Free Zone." Rumors circulated that the deal breaker for Myopic came when the producers indicated they would not film with anyone over age thirty in the store. Other business owners were more enthusiastic, few more so than Jack Wasserman, owner of the Local Grind Café. Located across the street in the Flat Iron Building, Local Grind quickly became the "Friends"-style hangout for cast members. Wasserman considered the show "great publicity," and said of the protests, "It would be devastating if viewers got the impression that we're an unfriendly city," adding that "the cast is really nice."[32] But the publicity did

Figure 6.3 Attempting to Keep "The Real World" at Bay.

not prevent Local Grind from following Urbus Orbis into oblivion within a year of the filming. Piece, a recently opened restaurant also across the street from the loft, experienced a surge of business when it hired two cast members as servers.[33] It continued to exploit this connection well after filming ended. The following spring, Piece made the airing of the show a weekly event at its premises, and a substantial crowd of fans viewed the show on multiple large-screen televisions there.

Artists like Gillam displayed their antipathy toward MTV by directing mild abuse (rude remarks and thrown paper) at the young cast members. This strategy fails to recognize structural similarities between the two groups. Like "The Real World" cast, artists in Wicker Park typically endure personal sacrifices in the hopes of culture industry fame, while receiving little in compensation from those who benefit most significantly. In this sense, *Real World* performers and neighborhood neo-bohemians are united by the fact of their self-exploitation. Some protesters seemed to grasp this fact; the funniest, and perhaps the most appropriate, leaflets posted around the neighborhood read, "Free 'The Real World' Six."

PART III

Artists as Useful Labor

CHAPTER 7

The Neighborhood in Cultural Production

I have no money, no resources, no hopes.
I am the happiest man alive. A year ago, six months ago,
I thought that I was an artist. I no longer think about it,
I am.
— Henry Miller, *The Tropic of Cancer*, 1934

THE ASPIRING ARTIST must be resourceful. As a film student at Chicago's School of the Art Institute, Delia financed her project in part with loft parties. Local deejays entertained gratis at these affairs, and completed scenes from the film-in-progress were projected against a brick wall. For the filming itself, locals in the café crowd were recruited to serve in various capacities — as set designers, camera operators, script consultants, or actors. Those lacking the requisite talents for one of the technically demanding tasks contributed as extras. Artists in Wicker Park participated in this sort of thing readily — after all, they expected the others to show up at the opening of their art exhibits, or the premier of their new play, or the read-through of their script in progress, or the finals of their poetry slam. A critical mass is necessary to support the production of culture for which there is as yet little popular demand or monetary support. It amounts to a bohemian bargain, and it helps us to understand why artistic sorts across genres continue to value urban propinquity.

In January 2000, I was an extra in Delia's film. The scene was filmed in a vacant storefront, made up to be an art gallery.

Figure 7.1 Flat Iron Studio Space.

The day before, at the Hardware Café where Delia worked, I received directions to the location, as well as instructions on the persona that I was expected to inhabit. In the film, a space alien arrives on Earth and, wandering through the streets of Manhattan, encounters various subcultures. In this scene, the alien is confronted by the spectacle of Lower East Side art mavens in action. I was to be one of those, and the chief characteristic I was instructed to perform was unbearable pretension. Since I went to gallery openings routinely and was a Ph.D. candidate, I felt up to the task, practiced at both art-show attendance and pretension.

Being an extra in a student film may sound like fun, and it is fun. However, on this occasion it was also very cold. The space that Delia procured had only a portable heater plugged into the corner. Nonetheless, some two dozen locals, mostly artists, gathered to shiver together for several hours. What was interesting is that my fellow extras, many of whom I had encountered at genuine openings, nonetheless felt obligated to play something other than themselves. They had shaven heads or exotic tattoos that were simply part of their everyday aesthetic, but on top of that they layered parodic costumes, over-the-top ensembles nonetheless culled from the reaches of their own closets.

One man was completely naked and covered in silver body paint. I thought that he must be very cold. I also thought that he must be playing the alien. He was not. It turned out that he was just meant to be an example of a New York artist trying to stand out from the crowd.

I said to Dan, who was standing next to me, that I had never actually seen a naked silver guy at any of the art events that I attended. Dan assured me that in New York, this would be the sort of thing I would see routinely. I find that claim implausible, but that is beside the point. In American popular culture, artists are routinely denigrated as effete elitists, naked emperors claiming superior style and intellect who are, in reality, simply pretentious and asinine (or silly). Apparently, these Chicago artists were both well aware of this caricature and happy to perpetuate it, while at the same time feeling personally exempt. Yes, artists are like that, they concede — *other* artists, such as those in New York. It is Chicago's revenge for its relative marginality in the world of popular and high cultural production, which is dominated by the coasts. Here Chicago's "second city" status is a source of comparative virtue — lacking glamour, Chicago compensates with unassuming authenticity. The stance echoes Nelson Algren's famous claim for the city: "Like loving a woman with a broken nose, you may well find lovelier lovelies. But never a lovely so real."[1]

On the other hand, one can be fairly certain that any gathering of moderate size in the Wicker Park scene will include some participants who will move to New York or Los Angeles within the year. Chicago turns out to be an especially congenial place for young artists just starting out, with ample opportunities to participate in off-Loop theater or the making of student films. Less culture industry concentration in Chicago also means the stakes are lower, allowing people room to experiment and generating a more nurturing atmosphere. And yet, many participants itch to make their eventual way into the big leagues, even if it means having to brush up against poseurs in silver body paint. Chicago, New York, Los Angeles, and a host of other cities are not so much in competition as they are differentiated nodes in a networked geography of cultural production, enmeshed in webs of exchange of both cultural products and human capital. Within large cities, particular districts emerge as privileged sites in this cultural economy, and Wicker Park is an exemplary case.

The Culture Industries

As is well known, culture is big business, with billions of dollars derived from film, television, and popular music. Though a great deal of ink has been spilled belying the banality (or worse) of culture industry products, in fact disseminating a steady stream of even mediocre cultural commodities demands harnessing a great deal of creative talent. Indeed, one thing that has never ceased to amaze me as I've gotten up close to pop-culture producers is just how much effort and individual virtuosity goes into the production of even formulaic crap.

The advent of television and the Internet, and the globalization of the film and music industries, have dramatically increased the demand for marketable cultural content. In the fine arts, which lack a mass market, strategies of profit valorization look more and more like the risk-intensive financial markets, and collectors speak of their collections as though they were investment portfolios (diversified living artists, some old masters, etc.), while dealers adopt the language of brokers. Still, as the art historian Thomas Crow indicates:

> In our image saturated present, the culture industry has demonstrated the ability to package and sell nearly every variety of desire imaginable, but because its ultimate logic is the strictly rational and utilitarian one of profit maximization, it is not able to invent the desires and sensibilities it exploits. In fact the emphasis on continual novelty basic to that industry runs counter to the need of every large enterprise for product standardization and economies of scale.[2]

At midcentury, Adorno and Horkheimer advanced the argument that the production of culture was managed like any other Fordist enterprise, within vertically integrated organizations following standardized routines.[3] But these industries have not merely grown larger in recent decades. They have substantially reorganized as well, providing excellent examples of the general economic trend away from Fordist mass production and toward "flexible specialization."[4] Fordist-style vertical integration now gives way to webs of flexible production.[5] In contrast to the old image of the culture industry worker as a studio-owned and -operated commodity,[6] the new trend is toward project-based labor markets[7] and chaotic careers for quasi-independent talent.[8] But while flexible production "liberates" labor from conventional organizational boundaries, this does not mean it democratizes the rewards. Media oligopolies and a select stratum of privileged cultural producers continue to dominate profits under new arrangements, while less formal relations of exchange with cultural producers in general absolve large but vertically disintegrated corporate concerns of bearing many associated market risks.

Under these circumstances, the commitment of contemporary bohemians to the romanticized images of starving artists and the primacy of the aesthetic does not confound the instrumental interests of the culture industries. Instead, the ideological features of bohemia work to the benefit of these industries, sustaining a pool of potential labor that largely bears its own costs of reproduction. Neo-bohemian neighborhoods help make this possible by clustering employment opportunities in areas like entertainment provision that help aspiring artists to subsidize their creative pursuits. The local ecology of neo-bohemia combines these opportunities with appropriate residential, work, and display spaces, creating a platform for artistic efforts that may then be mined by extra-local corporate interests, which recruit talent and co-opt cultural products from these settings at their discretion. This indeed did occur in Wicker Park during the 1990s; music industry scouts, for example, routinely scoured the neighborhood, signing many local acts to recording contracts. Likewise, the works of selected fine artists were exhibited in prestigious and profitable galleries on both sides of the Atlantic. In this case, it is useful to examine artists in somewhat unusual and counterintuitive terms, as workers in a cultural production process (rather than as, say, tortured geniuses or the heroes of modern life).

The Economic Profile of Bohemian Artists

Though current observers like Florida and Brooks prefer to stress the hedonistic dimensions of bohemia, in fact a durable feature of the artists' lifestyle has been the willingness to endure high levels of insecurity and material scarcity. Murger's tales of the Parisian bohemia depicted artists at the edge of destitution, often in failing health, conditions that by all accounts Murger himself knew intimately.[9] Since then, a poverty-stricken life has become part of bohemia's durable mythology.

This image bears scrutiny. Arguably, in the last half century the opportunities for artists to make a living have improved. In assessing what it means to live like an artist in the late twentieth century, it is important that we take a look at artists' economic profiles in the United States and, more specifically, in Chicago. To an unprecedented degree, the opportunity exists to earn large fortunes in both the mass media and the fine arts. This phenomenon has been enhanced by twentieth-century developments, as the economist Richard Caves notes, "One might suspect that superstars' careers were less starry in the past, if only due to inferior technologies of travel, communication, and the reproduction of creative works. In the 19th century neither the performer nor the performer's reputation traveled at today's speeds."[10]

Thus, though the artists of nineteenth-century bohemia no doubt harbored their own dreams of status and economic rewards beneath the "art for art's sake" veil, contemporary participants can have their efforts nurtured by fantasies of even more extravagant compensation.

Such an outcome is not, however, a common occurrence, and few aspirants in the arts will ever realize it. Pierre-Michel Menger reviewed the literature and concluded, "From a distributional perspective, artists show a high variance in income. Poverty rates among US artists are higher than for all other professional and technical workers."[11] The distribution of rewards is a classic example of a winner-takes-all market,[12] with a relatively small number of participants reaping the lion's share of monetary gains. Even for those fortunate few, the receipt of fame and fortune is often preceded by a de facto apprenticeship, perhaps lasting many years, during which remuneration for artistic efforts is intermittent and paltry.[13] Despite

Table 7.1

General Demographics for Chicago Artists by Percentage Income from Art, 2000 (number and percent)

Income from art (percentage)	No. respondents	With college degree or higher	Household income (annual)	
			Under 25K	25 to 40K
0 to 25%	562	85	26	29
26 to 50%	107	85	38	29
51 to 75%	28	91	39	36
76 to 100%	217	85	22	30

Source: Chicago Artists Survey 2000, Chicago Department of Cultural Affairs.

their relatively high levels of cultural and formal education capital, most artists find their creative pursuits significantly unrewarded by formal art markets of any kind.

Results of the 2000 Chicago Artists Survey, conducted by the Chicago Department of Cultural Affairs with more than 900 respondents, further illustrate this point. First, we must note the strikingly high rates of artists who have college educations; in total, 87 percent of respondents reported a college degree or higher. Still, in keeping with bohemian mythology of the starving artist, the return on this education in terms of household income is relatively low. Indeed, more than a third of these respondents reported household incomes of less than $25,000 a year. Seventy-three percent of them report making less than half of their income directly from art, and 61 percent report less than a quarter (table 7.1). Thus, most artists must support their art some other way.

Traditionally, bohemia is occupied by younger artists and fellow travelers, and Wicker Park is no exception. For Ephraim Mizruchi, bohemia is a "space of abeyance," in which participants forestall adult commitments; he suggests that the age of thirty is the tipping point beyond which such a slack existence ceases to be socially acceptable.[14] In Wicker Park, most of the more visible and active scene participants are in their twenties, many of them still students at local arts institutions, but a significant minority pursue la vie bohème into their thirties and beyond. Despite their age, these individuals continue to evince the styles and strategies typically associated with youth culture; they are more extreme examples of the American trend in which once standard markers of "adult" life, such as marriage and children, are increasingly deferred.[15]

In any event, the overall youthfulness of the scene means that local incomes are likely to be even lower in a neo-bohemia than the average for artists overall. Filer notes, "Although artists earn an average of 6 percent less than the general work force, the differential varies substantially over the life cycle. . . . Income differentials shift in favor of artists as workers grow older, so that above the age of 40 artists typically earn more than the control group."[16]

This "convergence," coupled with the fact that, overall, artists tend to be young,[17] suggests that many "failed" bohemians follow Mizruchi's advice and drop out.

The decline in public funding for the arts elevates income pressure for young artists. Federal funding sources such as the National Endowment of the Arts have been the object of significant controversy, decried by conservatives both for funding what they consider to be morally repugnant work, and for exemplifying wasteful government spending.[18] Federal funding for the arts peaked in 1981 and has declined since, although funding at the state and local levels has intermittently made up some of this shortfall.[19] Meanwhile, private funding for the arts via corporations tends to reward already established artists rather than aspirants.[20]

We can see, then, that at least in the early stages of their careers, contemporary artists do indeed earn meager incomes, especially when their high levels of education are taken into account. Still, the "poverty" of artists is of a distinctive flavor. Wicker Park artists like Delia are resourceful, after all. Even where direct economic compensation is scarce, cultural competence purchases local rewards, as young artists are routinely included in upscale events or staked to a drink or a dinner by more fortunate patrons.[21] Thus, starving artists do not really starve, and they often manage social lives far more lively than their earnings would indicate.

The bohemian relationship to material scarcity is complicated, not least by the personal insistence that this life was "chosen." Says Tom, a painter living on Chicago's West Side in the late 1990s, "I have something I do that is very important to me — making art — and I'm prepared to forgo other comforts for that. It's going to be a long time, if ever, before I own a new car — I never have. But that doesn't really enter into my consciousness." Such a noble claim serves to distinguish the artist from the great unwashed; as one Wicker Park entrepreneur put it, "There's a big difference in being 'poor by choice.' A bohemian who is poor by choice is socially acceptable. Someone who is poor because they can't earn any money is still déclassé." The neo-bohemian thus makes a claim on status privileges typically denied to the urban poor. While poverty as it is normally experienced inhibits self-determination, the "voluntary" adoption of relative poverty by bohemians claims to increase autonomy.

Bohemians may self-select into poor and working-class neighborhoods, but their dispositions are decidedly cosmopolitan, and they are not plagued by the social isolation that characterizes groups that occupy these spaces under conditions of considerably greater constraint.[22] Applying their creative talents to the construction of their own lives, they are practiced bricoleurs,[23] turning secondhand clothes into chic, trendsetting ensembles, and converting cultural capital into a myriad of social opportunities, from gallery openings to exclusive parties. Moreover, they are quite creative in reimagining the home and neighborhood that they occupy. Beyond low rents, these aspirants also require access to display venues in order to enhance their visibility in nascent careers; because their profiles are typically low during the bohemian phase, such venues must have relatively low barriers to entry, welcoming new and experimental work. Finally, the artists must be able to find work to subsidize pursuit of their art; ideally, they seek work that is consistent with their lifestyles. Wicker Park during the 1990s satisfied these requirements, though these various attributes of the milieu also often conflicted with one another, generating a precarious environment.

The Creative Milieu

In *The Informational City*, Manuel Castells advances the concept of the "milieu of innovation" to address new strategic sites in what he calls "the informational mode of production."[24] Writes Castells, "By a milieu of innovation we understand a specific set of relationships of production . . .

based on a social organization that by and large shares a work culture and instrumental goals aimed at generating new knowledge, new processes and new products."[25] Though Castells focused on enclaves of high-tech innovation, his concept is applicable to the production of culture as well, even if it challenges standard interpretations to consider the bohemian ethic a "work culture." When applied to neo-bohemia, Castells's emphasis on social interaction challenges the image of personal virtuosity that attends most representations of creative accomplishment. The cults that form around the most successful practitioners in the field obscure the complex social systems in place that actually generate cultural innovation.

Howard Becker argues that art is a cooperative venture: "All artistic work, like all human activity, involves the joint activity of a number, often a large number, of people. Through their cooperation, the art work that we eventually see or hear comes to be and continues to be."[26] The social world of cultural production privileges particular locales. Such places encourage the collective process of cultural production, fostering collaboration, linking artists to audiences, and sustaining a "work culture" through which participants come to frame their efforts. Such locales are not self-contained, bounded entities, but rather operate in multiple networks of exchange — of products, ideas, and human capital — with other key sites, preceding the eventual dissemination of selected cultural commodities into the global marketplace. These spaces attract relevant types of creative workers, such as artists, performers, and musicians, and they provide social conditions needed for the nurturing of talent and the ongoing creation of cultural products.

The Wicker Park bars and coffee shops that local artists, musicians, and hipsters claimed as there own, including Urbus Orbis, the Rainbo, Earwax, and the Borderline, are important for understanding more than just the recreational ethos of Wicker Park. In fact, such venues helped to organize a local scene that was home to substantial amounts of cultural creation, most notably in music, but also in the fine arts, theater, and design. Apparently a space of leisure, Urbus Orbis contributed to the real work of artists by promoting interactions and nurturing collaborations. It also created a general culture amenable to the arts, one in which an excess of definitions coalesced that supported bohemian values of self-sacrifice and the primacy of the aesthetic (in opposition to crass yuppie materialism).

Within the neighborhood, culture for profit may mean that an object of cultural production is sold on the market, like a painting, or it may mean that cultural objects or activities are used to sell something else, as in advertising. These dimensions may be simultaneously present — a local art fair may be designed to sell an artist's work, but at the same time, the fair may also be selling the idea of the neighborhood as a desirable place to live. Even activities like poetry readings that seem to have very limited economic potential still contribute to the creative ambiance of the neighborhood, and in so doing they increase its attractiveness both for cosmopolitan residents and aesthetically oriented enterprises in design and new technology. In either case, the practical activity of local artists generates value, even if someone else often ends up pocketing the profits.

In both the fine arts and more popular media like music and film, the cultural products that make their way into the cultural marketplace — what we actually see, what gets written about in the *New York Times* or even the *Chicago Reader* — represent success stories, and are only the tip of the iceberg in the process of cultural production. The existence of a vast cultural marketplace reflects an extensive selection process. For every cultural object that enters such rare air, an unknown number of actual or potential objects do not. But when we view the production of culture as a social process, rather than as a case of genius revealing itself, we can come to understand that these failures are a key part of the social condition that allows some to succeed. A dense community of young artists will necessarily contain only a handful who are destined for real success (measured by the standard indicators of media recognition and big dollars). When a large amount of cultural work becomes concentrated in a particular place, the job of gatekeepers responsible for discerning potentially marketable products is made easier. Communities like Wicker Park are like farm leagues, from which an unknown number of products and producers will ascend to the big leagues of marketable cultural production.

Particularly in the early 1990s, when its reputation achieved national proportions but gentrification processes had not yet dramatically raised local rents, the neighborhood contributed to the flourishing of young artists and musicians in the early stages of their careers, as well as a number of hip, funky hangouts where these artists could work and play. The media plays an important role in this process. When *Billboard* magazine creates a "buzz" about the Wicker Park music scene, it increases the future likelihood that music scouts, journalists, and art buyers will enter the neighborhood, which in turn makes it more advantageous for artistic aspirants who are hoping to be seen. The media thus simultaneously reflects and drives the development of neo-bohemia.[27] Despite the pretentious bandying of the word "underground," Wicker Park's postmodern bohemia is highly visible to a variety of relevant publics, from culture industry scouts to local consumers, and that is also part of what keeps the local artists coming back to the neighborhood.

Table 7.2 gives a brief catalogue of significant moments over the past thirteen years of Wicker Park as a cultural production site, revealing how local culture intersects with broader markets in music, film, and the fine arts. But this retrospective provides a necessarily fraudulent picture insofar as it renders static and discrete outcomes that are in fact embedded in a dynamic field of creativity and collaboration. The possibility for these outsize events relies on a host of more obscure efforts, a sea of cultural production that falls below the radar. Very little of this cultural production is valorized, or, more to the point, it is valorized indirectly, with large numbers of participants failing to find a market or reap the benefits of their contributions. Even after signing a band, the music industry seeks to limit its own exposure; as the Chicago-based independent record producer Steve Albini indicates, the record company advances promised to newly signed musicians often prove illusory even for moderately successful acts.[28] But in Wicker Park and similar neighborhoods, there is no shortage of musicians willing to take their chances, ignoring the fact that the risks they take are largely on behalf of corporate profitability. Other creative fields are similar in this regard.

Moreover, within the local milieu, compensations abound beyond the intrinsic satisfaction of the work. The starving artist, if successful at negotiating the status games of the neo-bohemian milieu, is compensated by an active social life and various perks of prestige that we can presume elude ordinary poor people. As Mizruchi indicates, bohemia constitutes

Table 7.2

Wicker Park and the Production of Culture, 1990–2001: Selected Events

9/9/1989	The first ever "Around the Coyote" (ATC) Art Festival in Wicker Park is held to increase the exposure of West Side artists; local businesses and artists' lofts serve as gallery spaces.
11/4/1991	New Crime Productions, founded by John Cusack and Steve Pink, opens Wicker Park's Chopin Theater on Division Street with *Fear and Loathing in Las Vegas;* features Jeremy Piven, future star of the ABC television series "Cupid."
6/8/1993	After seven years on the rock margins, Wicker Park band Urge Overkill makes its major label debut with *Saturation*, released by Geffen Records.
6/24/1993	Liz Pair's debut album, *Exile in Guyville*, is released on Matador; goes on to be named "Album of the Year" by numerous media sources, including *The Village Voice* and *Spin.*
1/22/1994	Filmmaker Rose Troche's debut, *Go Fish*, a 16 mm romantic comedy set in the lesbian community of Wicker Park, premieres at the Sundance Film Festival and becomes the first of the features there to gain a distributor. Made for $15,000, it grosses $2.5 million, while also winning the 1994 Berlin International Film Festival Award for Best Feature Film.
9/16/1994	Liz Phair appears on the cover of *Rolling Stone*, as part of its "Women in Rock" issue.
10/25/1994	The band Veruca Salt, a veteran of early performances at Phyllis' Musical Inn and other neighborhood venues, releases its debut album, *American Thighs*, on Geffen, spawning the radio and MTV hit single "Seether."
2/20/1995	Top-selling Chicago rock band The Smashing Pumpkins play the first of "four shows of the intimate kind" (announcement) at Wicker Park's Double Door, trying out material from their upcoming album *Mellon Collie and the Infinite Sadness.*
9/18/1997	Legendary rock idols The Rolling Stones play a top-secret show at Wicker Park's intimate performance venue the Double Door.
9/26/1997	The feature film *Soul Food*, directed by Wicker Park resident George Tillman Jr. is released to critical acclaim and financial success, inspiring a series on the cable network Showtime. Tillman goes on to direct the major studio release *Men of Honor* (2000) and to produce the smash hit *Barbershop* (2002).
3/31/2000	Under the rubric of New Crime Productions (now based in LA), the feature film *High Fidelity* is released. Pink takes co-writing and producing credits, and Cusack stars. Several scenes are shot in Wicker Park locations, including the Double Door. Domestic gross exceeds $27 million.

Table 7.2 (continued)
Wicker Park and the Production of Culture, 1990–2001: Selected Events

7/15/2001	MTV begins shooting its reality TV series "The Real World," using the Wicker Park loft at 1534 W. North Avenue as its principal location.
9/7/2002	The Dzine Wall Project opens at the Chicago Museum of Contemporary Art, showcasing the work of longtime Wicker Park graffiti and performance artist Dzine.

a tolerated lifestyle in which parents, friends, and potential lovers may forgive light pockets and shabby apartments since they come in the service of a higher calling that, as an added bonus, has the potential to suddenly generate extravagant wealth.

Wicker Park as a Training Ground

Becker acknowledges the existence of naïve artists, not exposed to mainstream conventions of art production, and folk artists who use conventions that are of a different and mostly unvalorized sort.[29] These more "organic" forms of cultural expression can influence commodified culture, but most of the culture that produces capitalist profit does not emanate from these sources. Some of the work of naïve and folk artists may make its way into the cultural conversation informing the more mainstream artistic media, although as it does it will surely be transformed by the transaction.[30] Most commodified culture is the product of people who think of themselves as serious culture workers in an established medium, and many of these people have formal training. Neo-bohemia concentrates these individuals. But it does not only contain creative human capital. The social experiences of the neo-bohemia, the "conversation" in which individuals have ongoing exposure to processes of cultural production by fellow practitioners, refines and enhances creative capital.

Most Chicago artists are well educated, often with formal training in their chosen field. Many participants in the local scene throughout the past decade are or were students at Chicago's School of the Art Institute or Columbia College,[31] which in 2000 combined to enroll more than 11,000 students in the South Loop area alone, with high rates of projected growth.[32] Formal arts education exposes apprentices to both technical training and an intensive regimen of art history and appreciation.[33] Other aspirants gain this knowledge through autodidactic efforts and informal dialogue. The neighborhood allows easy exposure to and discussion of a wide range of arts, contemporary and historical. Even creative aspirants who direct their efforts toward producing for low-caliber mass-culture outlets are likely to be well versed in the more canonical arts.

Sid Feldman speaks of the high levels of commitment that locals exhibited vis-à-vis their medium as the community took shape: "You had people who were very serious about what they were doing. . . . At the time, all the artists could talk at length about their medium, about its history." Populated by many such individuals, the bohemian milieu is thus a site for ongoing informal and eclectic training in artistic conventions. Alan Gugel, who received a fine arts degree at a major midwestern university, indicates that for him Chicago was a logical next step in his art education: I came to Chicago "to be involved in discussion and debate about art and society and what all this stuff is. So I came to the city because that's where it's at. That's where the discussion is, that's where it's going on." Cafés and bars like Urbus Orbis and the Rainbo Club provided the staging grounds for this sort of ongoing intellectual collaboration.

Harvey Molotch points out that "culture workers are typically each interdisciplinary and spark one another's energies across genres. The great majority of artists are active in more than one art."[34] Leaving aside normative judgments of cultural value, we can note that in fact new bohemias are characterized by the intersection of participants oriented to cultural forms that are obscure and unlikely to be valorized by the market under any circumstances (e.g., performance poetry) and to those that are intended to find a mass audience (e.g., popular music or film). Bernard Gendron has extensively documented the intersections between popular music and the avant-garde, from Montmartre cabarets like the Chat Noir to East Village dance halls like the Mudd Club.[35] One colorful example involves famous denizens of New York's 1980s bohemian scene; Madonna was a

stalwart of the East Village club scene and temporary paramour of bohemian icon Jean-Michel Basquiat before becoming one of the best-selling
pop musicians of all time.[36] In Wicker Park, Feldman devoured the works
of Camus and Genet even as he was actively strategizing to sell screenplays
for lowbrow caper comedies to Hollywood. The cross-fertilization of comparatively obscure cultural pursuits in the fine arts with the more popular
forms of the mass media is one way that bohemia acts as a generative field
useful to culture industry interests.

The stoking of creative impulses within the neighborhood also flows
from less explicit sources. Diversity and concomitant vitality are factors
of production in their own right that help us to understand the ongoing
association of urban districts with artistic activity. Benjamin provocatively
examined the relationship between the observations of the urban stroller
and creative dispositions, resolved in the figure of the *flâneur*.[37] Central to
the *flâneur* is the diversity of the city street; the *flâneur* encountered not
only the dandy on his walks, but also the ragpicker.[38] In the city, social
heterogeneity contributes to what Georg Simmel calls "a mental predomi-
nance through the intensification of consciousness,"[39] and this intensification of mental life underlies modernist innovation.

The pedestrian on the sidewalk remains a key figure in understanding
the magical nature of urban life, juxtaposed to "the virgin sidewalks" of
suburbia; it is pedestrian life that both Jane Jacobs[40] and Richard Sennett[41]
write about so compellingly in their respective meditations on Greenwich
Village. In Wicker Park, artists continue to speak to the importance of
diverse sidewalk life. One describes it as "the difference between having a
culture and not having a culture. Culture is you're walking down the street
and you see a poster, and you read it, it looks interesting to you, and you
go to see it." Adds another, "I like the beat of the city. The pace. There are
so many things happening in the city on any given thirty seconds that you
can add to a story, and you can either use those — they could be some kind
of symbol, or hey, maybe they'll just be there. . . . I believe you can paint
a scene just with these little glimpses that a person might see as they're
standing on a corner waiting for a cab."

Such sentiments are direct descendants of the impressionist street scene,
or Baudelaire's urban poetry.

Exposure

Wicker Park artists with whom I spoke often indicated that although career upside is more limited in Chicago compared to New York or Los Angeles, it is much easier to get started there. Shappy graduated with a degree in theater from Eastern Michigan University in 1991 and came immediately to Chicago and Wicker Park. He recalls:

> I knew a lot of people that had already moved to Chicago and told me it was really easy to start your own shows and stuff — which it was. I couldn't believe how easy it was for us to get a space. With all the creative people . . . it was very easy to collaborate, to come up with scripts. I remember we were writing and producing our own material like the minute I got into town. I went to all the open mikes, tried to do some comedy, I slammed [competed in a poetry competition] in '91 or '92, I think, when I first moved here.

Along with his collaborators, Shappy almost immediately mounted a sketch comedy show called "Every Speck of Dust that Falls to Earth Really Does Make the Whole Planet Heavier," staged in the upstairs of Urbus Orbis before it converted into a futon outlet. Over the years he developed a substantial local and national following for his poetry and comedy, hosting comedic "Shappenings" at venues like the Note and the Empty Bottle, and traveling with the heavily sponsored summer music festival Lollapolooza as part of the poetry tent. In 2001, after many years honing his manic persona, he moved to New York.

Steve Pink, whose New Crime Theater Company opened the Chopin Theater in Wicker Park with a production of *Fear and Loathing in Las Vegas,* moved to Los Angeles in 1994. This veteran of small Chicago theater now works developing film and television projects, co-writing and co-producing the major studio releases *Grosse Pointe Blank* (1997) and *High Fidelity* (2000), both starring Pink's Evanston High School classmate John Cusack. According to Pink, despite the enormous number of writers and performers that concentrate around LA's film and television industries, the scene there is comparatively atomized, undermined perhaps by the intense competition for film and television dollars:

There is culture [in LA], but there is no interaction, there's no interconnectedness, there's no people being expressive. When we were doing theater [in Chicago], everyone came and saw our shows, and liked them or criticized them, and then we saw everybody's shows, and everyone knew what we were doing. I'd see people from 20 theater companies every week, you know, you don't see that [in LA]. In LA, "the Industry" is film and television. . . . That's what people do. On a much higher kind of financial and class level, that's all happening. I don't go to the parties. They bore me. 'Cause it's not interesting. You know what I mean, its about making films and making money and being successful and making deals, so it's a lot different. . . . The difference is in LA everyone thinks you're doing it for the money, even if you're not, it doesn't matter . . . because the money is involved. So it kinda sucks. There's a kinda dirty sordid aspect to it, which is: you're doing it because you want to make money, you want to be successful. You're not doing it because you want to express something specific. If you even say you want to express something specific, and you say it with such pure idealism, that it doesn't really occur to you that you'll make money doing it, then people look at you like you're insane, and you sound insane.

The lower stakes in Chicago make it a better spot for the development of new talent that can then, like Pink, relocate, and for the trying-out of more innovative products. Chicago theater is a staging ground for New York as well as LA, with theatrical hits like *The Producers, The Sweet Smell of Success,* and Brian Dennehy's turn as Willie Loman in *Death of a Salesman* all first running in Chicago, working out their kinks away from the big-money pressure cooker of Broadway. Stephen Kinzer reports in the *New York Times*:

> Robert Brustein, artistic director of the American Repertory Theater in Cambridge, Mass., said Chicago had become "the city that's of primary interest to theater people in New York. It's a big city, funding channels are extremely generous, and there's a nice electricity there that vibrates into culture. . . . Chicago does something that

New York isn't doing much anymore, which is taking chances on new plays. There's so much financial pressure on New York theaters these days that they feel unable to take those chances."[42]

Though Pink now lives full time in Los Angeles, he retains ties to Wicker Park and to Chicago. He successfully lobbied to have *High Fidelity* filmed in Chicago, despite the fact that the Nick Hornby novel on which it is based was set in London.[43] The film tracks the romantic mishaps of Rob (Cusack), a hip, disaffected slacker and owner of a secondhand record store called Vintage Vinyl. The store in the film is in the heart of Wicker Park, where much of the shooting took place. Says Pink, "Rob's record store was on Milwaukee Avenue, across the street from Burger King, across the street from Salvation Army, because that's where alternative records stores are, and that's where alternative culture exists. It existed for us. We needed him to walk those streets." Indeed, the actual alternative record store Reckless Records is only a hop, skip, and a jump away from Vintage Vinyl's fictional location. The climactic scene of the film is set a few blocks away on Milwaukee, at the real life Double Door, where Rob hosts a release party for a recording by a band of local skateboarders that he produced. The character's personal growth is signaled by his move from being a cultural curator to a culture producer, drawing on local resources.

Pink continues to extol the virtues of Chicago as a generative milieu for cultural production, noting that the city offers him more in the way of inspiration as a writer:

It's better in Chicago, because you read the paper and talk to people and you're exposed to things all the time. So much better in every single way. You can see a hundred thousand people a day in Chicago if you went downtown. In California, you don't really see them; you just see their cars. You're not exposed to things happening. You're isolated. That's the difference. In some sense, really all I'm saying is it's more fun to write in Chicago because I'm more inspired.

To be sure, many people, including Pink, do write for a living in Los Angeles. In addition to the thousands of hopeful or actual film and television writers, Los Angeles is also a seedbed for one of the most distinctive literary traditions in U.S. history, noir, exemplified in the work of James M. Cain, Raymond Chandler, and, more recently, James Ellroy. In *City of Quartz*,[44] Mike Davis offers an excellent analysis of noir as a distinctly LA art form. But as Davis also points out, there is a long history of ambivalence toward the LA effect on creativity by the artists and writers who have worked there, and this too is registered in cultural products, from Nathaniel West's novel *The Day of the Locust* (1933) to the Coen brothers' film *Barton Fink* (1991).

There is no one way that place contributes to the process of cultural production. Sites of cultural production as disparate in character as Wicker Park and Hollywood are united by networks of exchange in a multi-site division of cultural labor. Throughout the 1990s, Wicker Park evinced properties that encouraged an influx of individuals working in creative media, and its reputation helped to link these aspirants to relevant industries based in cities like New York and Los Angeles, as well as closer to home in Chicago's Loop.

Filling the Room

Though the cultural knowledge and technical skills that abound among Wicker Park artists only rarely lead to their individual market success, the artists have beneficial effects on the local art market as a whole. Individuals that have invested time and energy into mastering complex cultural conventions become exceptional consumers of culture. They are positioned to get in early on other cultural innovations, to value and reward new artistic efforts before a wider audience does this. Indeed, participants pride themselves on their ability to do so, and the constitution of bohemia involves a commitment by members to being ahead of the game, on the cutting edge. This local selection process raises the visibility of some artists, making the work easier for more formal cultural gatekeepers on the lookout for potentially marketable cultural objects. Other consumers, who are also attracted to some notion of the cutting edge, use the consumption practices of artists as a model as well, and thus the number of participants in such a scene swells.

Thus, for most of the cultural offerings in the neighborhood and near-by, including musical acts, poetry readings, and art openings, a substantial portion of the audience will be artists themselves, although not necessarily artists who work in the same medium. As Pink indicated above, a collective bargain is maintained, essentially that "I'll go to your shows if you come to mine." Indeed, William Bullion, a founder of Sliced Bread Productions, tagged the e-mail announcement of a 2001 play that he directed with the plaintive reminder "I go to your shows." Art openings, musical perfor-mances, and theatrical productions fill the social calendars of local artists. Particularly strong new cultural offerings will generate local buzz among the creative community, another factor influencing the eventual selection process by formal industry gatekeepers.

Even when not spending money, artists contribute to local strategies designed to turn cultural production into profit. Paul Klein, the owner of a successful gallery in River West, estimates that at an opening, only 10 percent of those who attend actually have the potential to purchase a work on display. But it is important to "fill the room" he says; if attendance is slack, it "looks like there's not enough interest in the work." Artists are likely fillers, projecting an appropriate look and demeanor for such events gleaned through iterated participation. They decorate the room, and are available to discuss the work with potential buyers in an apparently knowl-edgeable way, conveying enthusiasm for the arts in general. Klein says, "At an opening, a small number of people have the potential to buy, a fair number are artists, and a fair number of people just want to check out what's going on and have some connectivity to the art scene. Categorically, I'm interested in [attracting people] who care about art." For a commercial gallery, potential buyers provide the raison d'être, but the numerical major-ity of artists and enthusiasts provide the ethos. In return for their efforts, these art-opening pros get to chart a cosmopolitan social calendar, chatting with financial elites while being nourished by free wine and cheese.

Liz Phair

Liz Phair, whose debut album *Exile in Guyville* reflected an ambivalent re-lationship to the neighborhood and its highly male rock ethos, generated both media adulation and local resentment when she burst onto the music scene in 1993. Though she was deeply involved with the thick local scene prior to the release and subsequent success of *Guyville*, Phair insists that her creative talents were not taken seriously by her male counterparts.[45] Indeed, the local painter Tom Billings remembers her as being an "artslut" rather than an artist prior to the release of *Guyville*. Wrote Billings in *SubNation*, "An artslut is someone who works for other artists, preparing canvas, woodburning, basically doing the work that bores other artists."[46] Though many local artists knew Phair as a regular at the Rainbo and other bars, most remembered her as being "no big deal" artistically before *Guyville*'s release; many did not even know she was a musician. Recalls Phair of the male indie rockers in Chicago, "[They] always dominated the stereo like it was their music. They'd talk about it, and I would just sit on the sidelines. Until finally I just thought, 'Fuck it. I'm gonna record my songs and kick their asses.'"[47]

On the other hand, Phair did manage to enlist the talented local producer Brad Wood to record *Guyville* (and lay down its drum tracks) in his Ideful Studios in Wicker Park, and she used local musicians as backup in live performances. Her first live gig took place at Czar Bar, next door to Phyllis' Musical Inn. So despite her feelings of personal rejection, she still took advantage of local opportunities for collaboration and display. Further, as this case suggests, competition as well as cooperation can be a way that the scene inspires the efforts of some creative aspirants. Phair blasted the ego-driven ethos of the indie rock scene, but interviews following her success demonstrated that she herself had no deficit of self-regard, despite the shyness evidenced by her legendary stage fright.

Taking frequent refuge behind a façade of sexual bravado ("I take full advantage of every man I meet"), Phair describes in her music a world of one-night stands and petty humiliations, as well as her aspirations to show up her antagonists on her own terms.

Phair did indeed weave her reflections on "guyville" into fame, although her success hardly endeared her to all of her local peers. The *Village Voice*, based on a survey of roughly 300 critics, named the then–26-year-old rocker "Artist of the Year." Incredibly, she was the first woman to be so designated by the *Voice* since Joni Mitchell in 1974.[48] In 1994, she graced the cover of *Rolling Stone* magazine.[49] Perhaps shamed, surely jealous, many Chicago artists continued to regard Phair as a lightweight and a sellout, in contrast to the critical respect afforded to her outside the neighborhood. Indie record producer Steve Albini dismissed Phair as "a rich suburban girl who made a name for herself (by) having an incredibly aggressive marketing campaign come to bear."[50] But Phair, more than anyone else, put the Wicker Park music scene on the national map. Says a local musician, "I think she opened the door, and I'm glad. I'm glad somebody had the balls to do that, even if it was Liz with her balls." This characterization of Phair is especially telling, given that the music press has given her a great deal of credit for providing female rock musicians with more "industry validity," helping to pave the way for a rash of subsequent signings of artists including Lisa Loeb, Fiona Apple, and the true breakout of the bunch, Alanis Morrisette.[51]

It is clear that Phair alienated some locals by making her ambitions too plain. For all its critical success, the challenging *Guyville* was a solid but not huge seller, partly because its sexually explicit lyrics made it difficult to back with radio airplay.[52] After making a six-figure deal with Matador, a nominally independent label distributed by the much larger Atlantic,[53] Phair made an effort to be more commercial, and she was not shy about admitting this, even suggesting in interviews that her "cuteness" could be a selling point: "I'm cute enough that you can photograph me, you can dress me up, and I'll do it, I'll smile and dance around."[54] She may have misrecognized the appropriate pose for an ambitious artist in the "alternative" musical genre popular in the 1990s. Despite *Rolling Stone's* cover declaration that "a rock & roll star is born," her sophomore effort *Whipsmart* never cracked the *Billboard* top 100, selling in the solid but unspectacular range of about 250,000 copies.

Further, even many of her supporters were disenchanted by its more commercial tone. However, her third album, *whitechocolatespaceegg*, dealing with adult themes of marriage and parenthood, was better received by critics. Meanwhile, she augmented her income with modeling, appearing on a lower Manhattan billboard for Calvin Klein, and by contributing songs to various popular films (*13 Going on 30*, *How to Deal*, *She's All That*, *Chasing Amy*) and television shows, including MTV's "The Real World: Chicago." With the song "Shitloads of Money" she responded to critics for her material girl program, arguing that while "most of her friends" espoused conventional bohemian antipathy towards the market, this stance concealed a secret longing for success, including, well, shitloads of money.

Critics were especially rough on her 2003 release, the eponymous *Liz Phair*, in which she enlisted the help of the production team The Matrix, best known for their work with bubblegum punk purveyor Avril Lavigne. In this case, the consensus seemed to be that Phair had substituted a highly packaged, faux subcultural style for her rawer and therefore authentic earlier efforts. In 2004, I hit San Francisco's legendary Fillmore Theater to watch Phair headline the "Chicks with Attitudes" tour (also featuring veteran Swedish performers The Cardigans and the teen-aged chanteuse Katy Rose). Clad in apparel appropriate to her latest bout of reinvention, if not her age, Phair had shed the awkwardness on stage that characterized her Wicker Park-era outings. Her dancing was energetic and self-assured, and years of voice work since her debut had immensely strengthened her vocal delivery. Still, despite her efforts to court the teen pop audience, the crowd in the all-ages show was heavily populated by patrons who appeared to be, like Liz and me, in their thirties or beyond. Nor did she neglect those in attendance who were attached to her earlier efforts; indeed virtually the entirety of *Guyville* was included in her set. And it was the songs from her breakout debut that routinely elicited the most enthusiastic response from the audience, including the two teenage girls standing near me on the floor that sang along word for word with "Fuck and Run" and "Flower."

Prior to her success, Phair was already heavily invested in the neighborhood's internal status hierarchy. She confessed in later interviews, "It was important to me to be hanging out with the coolest artists," some of whom she was linked with romantically. But Phair was not herself highly placed in that hierarchy (as Billings' "artslut" characterization indicates), demonstrating the limitations of the local status system as a predictor of future success. And even the massive critical success of *Guyville* did not lead to a universal reevaluation on the part of locals. Her parting with the neighborhood was not without acrimony: "There's a lot of suspicion and bitterness when someone does make it," Phair told *Billboard* in 1993. "The standoffishness of the indie scene just screams insecurity to me."

Local Status Distinctions

As Phair indicates, the artistic milieu is characterized by insecurity, along with associated petty jealousies and cliques. Tensions were exacerbated in Wicker Park as the art scene attracted more attention and some participants achieved fame. However, even prior to this the milieu was not egalitarian, despite one artist's claim: "There were no stars then, just a bunch of dumb artists hanging out." Some in the neighborhood were the "coolest artists" before anyone got famous, and these were not necessarily the ones who got famous. The local scene, like other urban bohemias, provides a setting in which issues of self-understanding and notions about what "being an artist is" are worked out, usually far in advance of what Tom Wolfe calls the "consummation" of material success.[56] Local status is easier to come by than meaningful market rewards, but it is still earned in a competitive field, in which individuals are ranked in a hierarchical system of distinction.

The competition for local prestige accelerated as the increased visibility of Wicker Park's arts scene in the local and national press brought with it new aspirants clamoring for participation. This development led to fine distinctions made on length of residence, in which local artists already living in the neighborhood (for whatever length of time) would catalogue the comparative inadequacy of newcomers. Among the complaints lodged against the increasing numbers of young people that formed a second and third wave of bohemian aspirants was that they were more interested in making the scene than they were in making art. Feldman suggests that there was a qualitative difference between the artists of the late 1980s and early 1990s, many of whom graduated to successful careers, and the next generation: "All these people that were around here were very committed to their art, and very good at it, and some of that's changed. A lot of people started moving in from the suburbs and saying, 'Yeah, I'm an artist! Wooh!' And they're not so committed. They're not so educated about what they're doing." Jimmy Garbe expresses a similar sentiment: "It [used to be] a different breed of artist. They were, to me, a lot more serious about their art. They're like lifers and really, really dedicated, compared to what I see now. Anybody who picks up a brush or a guitar labels themselves a musician or an artist, and to me

I don't think everybody's putting their heart and soul into it. I think it's become as much of a hobby as a career move or passion."

Thus the paradox of urban hipsters seeking to occupy the cutting edge, expressed by one San Francisco–based musician I spoke to: "They want to be out in front, but not so far that no one can see them." For many in the neighborhood, the explosion of media popularity and the arrival of next-wave artists ratified what they had already suspected, that they were onto something special. Some used the neighborhood as a springboard to fame and material success, but many others were disappointed to find that they were simply priced out of their lofts. Still, even if they are destined to go uncompensated, these individuals have played their part in forging the Wicker Park milieu of innovation. The value of cultural commodities is the result of a social production process, but the profits generated by this activity tend to concentrate in a tiny number of hands. In Wicker Park, a large number of aspirants not only are alienated from the profits, but must scramble to subsidize their own exploitation.

CHAPTER 8

Making the Scene

In a real dark night of the soul,
it's always three o'clock in the morning,
day after day.
— F. Scott Fitzgerald, "The Crack-Up," 1945

In HIS PIONEERING 1970S BOOK *The Tourist*, Dean McCannell writes, "The industrial epoch has biased sociology in several ways. Our research is concentrated on work, not leisure, and on the working class, not the middle class."[1] A perception of what constitutes "real" work, formed during the industrial period, persists stubbornly among social scientists, privileging blue-collar manual labor as the proper focus of serious study. But leisure activities and middle-class consumption also imply processes of work and social production. With the hollowing out of the old industrial core, service provision becomes a major source of urban employment. The educated professionals who constitute the elite sectors of the postindustrial economy demand a distinctive amenity mix in the city, raising the importance of what Zukin calls the "symbolic economy."[2] In Chicago, culture and entertainment are now core enterprises, both in terms of exportable cultural products and local entertainment scenes.[3] But the requirements for such production remain poorly understood.

The "industry" jobs in the bars and restaurants of the West Side's neo-bohemia differ from the bulk of the work performed in food and beverage retail. The uniformed employees at fast-food restaurants manufacture practically identical hamburgers as if on an assembly line, following formally rational procedures to crank out standardized Big Macs.[4] Writes Zukin, "In contrast to the depiction of the labor process as flexible, self-propelling, and intellectually demanding, many service industries rely on extreme standardization of labor, multiple levels of managerial authority, and rote performance."[5] In these jobs, principles of Fordist production persist, although absent the rising wages and job security that characterize the Fordist social contract of capitalism's "golden age." As Katherine Newman shows, this kind of labor increasingly becomes the primary source of (legitimate) employment opportunity available for those inner-city populations whose life chances have been almost decimated by the flight of traditional manufacturing jobs from the city.[6]

Conversely, managers and owners in Wicker Park's trendy bars, restaurants, and nightclubs do not provide a routinized script for their employees to follow. As in many other post-Fordist workplaces, the individual creativity of employees is a thing to exploit, not to suppress.[7] In Wicker Park, uniforms are a rarity, and where they exist they are juxtaposed to the worker's striking individual characteristics. Instead, the workplace norms of appearance usually require that workers display a honed neo-bohemian fashion sense — at their own expense — in order to improve the overall ambiance of hipness. In addition to their aesthetic contributions, such workers are also expected to put their personalities on display, with their mastery of nocturnal comportment serving as a blueprint of cool for scene consumers.[8] Employers purchase the aesthetic predilections of young artists, nurtured in the local subculture.

Arlie Hochschild notes that the service industry demands emotional work, and she documents the way in which flight attendants, expected to produce smiles that aren't just painted on, experience alienation from their own emotions.[9] Service workers in Wicker Park also contribute emotion work; additionally, artists are likely to incorporate highly aesthetic self-presentation into their persona, along with the argot and demeanor of the urban hipster. This self-presentation satisfies the expectations of young professional patrons, who may also have read *On the Road* in college.

In his study of Chicago Blues clubs, David Grazian dissects the construction of a nocturnal self: "That is, a special kind of presentation of self associated with consuming urban nightlife."[10] Construction of this persona draws on a range of competences, such as fashionability, that Grazian refers to as "nocturnal capital." It goes without saying that some possess more of this sort of competence than others, and the bohemian milieu is a setting in which a cool demeanor and hip fashion sense is nurtured. Moreover, the neighborhood is a place where these competences are convertible into economic value, value often captured by players other than the artists themselves. Thus, aesthetic self-work on the part of young service workers "makes the scene" in Wicker Park and may be highly valued by discerning consumers. Entrepreneurs strategize to take advantage of these attributes, going out of their way to hire individuals whose exotic and funky personae elevate them to the status of attractions, or "bar stars."

How Middle-Class Kids Get Working-Class Jobs

But why do young artists accept work that is physically demanding and possibly demeaning, making a mockery of their often superior education and middle-class pedigrees? To recast Paul Willis' question slightly,[11] Why do middle-class young people take working-class jobs, all the while defining the situation such that it is not experienced as downward social mobility? In Willis' study of working-class youth in 1970s Manchester, England, he argues that the counter-school subculture of the tough kids is rehearsal for a life of manual labor, its participants learning to identify with their brawn rather than their brains. The irony is that a strategy of "resistance" ends up contributing to the reproduction of their own economic marginality. The contemporary bohemia likewise involves ostensible resistance to the terms of bureaucratized, "corporate" labor-force participation. Ironically, such resistance now contributes to the ready acceptance by educated intellectuals of manual-labor jobs in the post-Fordist entertainment industry.

The manual labor of service industry work in Wicker Park is very different from that found on the manufacturing and construction job sites that eventually employ Willis' lads. Though labor in the bars and restaurants of the West Side is often physically grueling, it also requires varying degrees of performative competence, that is, the mastery of hip social codes. The anti-intellectual ethos engendered among Willis's lads does not serve these labor requirements well, but the countercorporate culture of neo-bohemia does. Neo-bohemia nurtures both the aesthetic competence and the subjective dispositions required by these jobs.

So unlike Willis' account, this is not a story of social-class reproduction. Taken on their face, these jobs appear to represent downward class movement, but this is also not an entirely satisfying interpretation. Participants would certainly resist it, and the contingent and partial nature of their service employment makes it hard to locate in a conventional class framework. Typically, service sector jobs are considered to be temporary stops on the way to something else — few artists take them with the idea that they will last more than a few years, and the premium on youth in the bar scene discourages extended terms of employment in any event. Nonetheless, many of the service workers/artists that I encountered were in their mid- to late twenties or early thirties — that is, well into adulthood. There can be little doubt that for some, the self-designation of artist amounts mostly to a self-legitimization strategy through which the children of the middle class can experience servile labor as something other than downward social mobility. Moreover, many serious young artists discover over time that the demands of service employment and its associated lifestyle are less compatible with their artistic interests than they initially thought. Since the odds are stacked against any given artistic career taking off financially, the question of what these jobs are a stop on the way *to* begins to loom large.

An art degree nearing a decade old and a résumé of bartender and wait-staff jobs is not likely to entice a wide range of professional employers. Nor is the laboring habitus engendered through years of service work compatible with the many occupations whose norms of dress and comportment — and working hours — differ substantially. Like other manual labor jobs, service work takes its toll on young bodies, and the damage is compounded by the high incidence of substance abuse in the industry. As Krystal Ashe, a local bartender and writer, observes, "It's easy to get sucked in, and I've seen a lot of people lose their looks, quit college."

Artists take service sector jobs on the premise that these jobs will support their "real" work as cultural producers. Says Krystal, "I'm not bartending because I can't get another job or because I'm an alcoholic and I need to be around drugs or anything like that. I make it work for me. I can spend time writing something if I want." After all, work schedules are typically less than full time, often as little as twenty hours a week, with significant flexibility to rearrange shifts from week to week. For the most part, it is nighttime work, seemingly leaving days free. But as Amy Teri, a former Borderline bartender, notes, "You get your days off, but you get home at five in the morning and you sleep all day."

Young people in the service industry typically find that it encroaches on their lives to an extent disproportionate to formal work hours. Nocturnal dispositions and embeddedness in a thick network made up of other local artists/service workers produces social pressures to spend an inordinate amount of personal time in the same sort of bars as those where they toil, pressures that many service workers are unable to resist. The labor of bar work, along with the associated party life, proves extremely draining, and thus may be incompatible with the diligent pursuit of other interests. Amy Teri was still employed at Borderline when she told me:

> I think it happens with a lot of people, they get stuck in the hustle-bustle of this job, and they quit doing the things that they were inspired to do, such as their artwork. They're caught up in the bullshit and they're too tired to explore their outlets any more. . . . This job can be very distracting if you're trying to do it for a living. . . . I think it has a huge impact on people evolving. I think people really get stuck.

She added, "If I'm still here in a year, kill me."

Because personal style is a high priority, young artists also value these jobs for the freedom that they provide to cultivate one's "own look" and cultural tastes. In Wicker Park, many artistic aspirants are more successful producing themselves as a work of art than they are at producing art itself, and the bar/restaurant scene valorizes these efforts. Says Amy Teri, "I guess that the good thing is that you can you listen to your own music, and you don't have to wear a uniform." Katie Baker, a Borderline bartender with aspirations as a clothing designer, says, "I usually dress up [at work]. I definitely use it [the job], a lot of the time, to wear my clothing and use it as my own little advertising nook. . . . I can wear what I want. I can listen to my music. I can be myself." Such sentiments highlight the differences that neighborhood service workers experience in their jobs from more Disneyfied service employment. This expanded "freedom" generally suits neighborhood employers, who incorporate the hip tastes of employees into their venues' aesthetic.

Broadly speaking, working as a waiter or a bartender is not an occupation traditionally conferring much prestige.[12] The sociospatial field of neo-bohemia has its own status norms, however, and artists as service workers have an outsize amount of local prestige, especially the workers in the better-regarded establishments. These jobs put their occupants on stage in the seemingly glamorous world of hip entertainment provision, and their ability to provide favors such as expedited service and the occasional free drink make them sought-after friends. In a study of fashion models and new media workers, Gina Neff, Elizabeth Wissinger, and Sharon Zukin argue that these cultural industries are marked by "'cool' jobs, 'autonomous' workers, and fluid organizational structures — at a cost of job instability, foreshortened careers, and winner-take-all rewards."[13] Cool jobs generate local prestige. The reward structure of bar and restaurant work is more circumscribed than those of Neff's targeted occupations, but the imagination of "coolness" and "autonomy" clearly applies to service workers' understanding of their occupations' status appeal. The costs of job instability and foreshortened careers are likewise the salient trade-off for a cool bar job.

Playing the Game

When reduced to the basic tasks required, the activities of preparing and serving drinks or waiting on tables are highly repetitive and not particularly demanding on the intellectual capacities of service workers. At the same time, these jobs are physically strenuous, especially on high-volume nights. Aside from traditionally low pay, it is the relatively low demands on abstract reasoning capacities and high demands on physical labor that have led these jobs to be regarded as working class and held in low social regard. This is complicated in the case of artists as service workers in the hip bars of the West Side. But though artists distinguish themselves with their aesthetic contributions, this does not absolve them of the ordinary, unspectacular obligations that go with the job. As Krystal Ashe notes, "Bars generally hire [people] because of their looks. They will hire people because they look good that don't know how to do their job. [But] once you get in the door you start working, and nobody is going to keep you if you can't make drinks."

Making drinks is not complicated. Although my 1988 version of *The Mr. Boston Official Bartender's Guide* (the venerable gold standard of bar books) lists several hundred drink recipes, bartenders in real work situations rarely encounter more than a few staple orders, many of which, like a gin and tonic, have names that are self-explanatory. Patrons who attempt to order complicated mixed drinks with puerile names like Sex on the Beach or a Screaming Orgasm are automatically typed by bartenders as hopeless amateurs and suburbanites, and at bars like the Borderline, these requests are summarily refused. "This is a drinking bar," says Amy Teri. "This is beer and hard liquor. I won't make those 'foo-foo' drinks." Though bartenders in Wicker Park experience themselves as actively constituting the establishment's aesthetic, they do not see their creative contributions as residing in the ability to mix terrific Bloody Marys.

The challenge of bar and restaurant work is not in amassing an extensive repertoire of drink-making skills, but rather in being able to respond quickly to the demands of the customers, and to make rapid-fire decisions about what must be prioritized at a given moment among a number of necessary tasks. This comes with experience, accomplished once the simple tasks of locating the right bottle at the right time become rote. Bartenders are constantly on their feet, and during rush periods, which might on a Friday or Saturday night last for several hours, they are in almost constant motion. Since the demand can be literally relentless during crowded periods, the variance in profit becomes entirely a function of the bartender's speed and efficiency in dispensing drinks. Whatever other contributions an artist may bring to the scene, the owners of these establishments will not tolerate a bartender who either cannot or will not keep up the pace during these grueling periods.

For all the differences between service sector labor and industrial manufacturing, the basic requirement that employees' labor generate profit over their wages still applies. Studying factory workers, Burawoy poses the question: "How are workers persuaded to participate in the pursuit of profit?"[14] Likewise, we might ask how management gets service workers in Wicker Park to work as hard as they do satisfying the demands of patrons, in the process creating the fundamental conditions of a bar's profitability. Burawoy writes of the efforts by employers to secure diligent labor, "Obviously one way — but generally not a very efficient one — is by continuous application of coercion, that is, by firing those who do not achieve a given quota. Coercion, of course, always lies at the back of any employment relationship, but the erection of a game provides the conditions in which the organization of active cooperation and consent prevails."[15]

Employers in Wicker Park certainly have recourse to coercive tactics; firing of service employees is common. But as we have seen, owners and managers (often the same people), prefer to employ other strategies.

The willingness of service employees to really sell out, delivering exceptional efforts in the face of extreme demands, is not a function of the low base pay they receive, which industry workers indicate is a negligible part of their total compensation. Nor does it spring from their exceptional loyalty to the entrepreneurs who employ them; despite the communitarian efforts of employers like Landise, service workers around the neighborhood most often express antipathy toward their bosses. Loyalty toward other employees appears to provide stronger incentive. The limited space behind a bar generally means that no more than three bartenders can be working at any one time, for example, and should one fail to keep up the pace, this is likely to create more work for the others and engender resentment. Under heavy assault from demanding patrons, a foxhole mentality no doubt develops. But these concerns are dwarfed by the motivation inherent in the real source of compensation for bartenders and servers: customer tips.

Tips serve a function similar to the piece rates paid to Burawoy's factory workers, as a variable source of income, the pursuit of which gets constructed as a "game" by employees. Though this may appear to not be a management strategy at all, but rather a discretionary reward supplied by customers themselves, in fact owners forgo some potential income and cede a certain level of workplace control so that workers can strive to maximize tips, in the process both increasing workers' personal income and allowing them to play an absorbing game.

The ability to generate substantial tipped income is enhanced by the willingness of bar owners to endure a "shrink," that is, a certain number of drinks given away, mostly to other local service workers and neighborhood hipsters. For many establishments, the arrangement is explicit. Matt Gans, who managed West Side nightclubs like the Funky Buddha Lounge and Wicker Park's Big Wig as well as the restaurant Mirai, told me, "You set up guidelines. Like 'Here's your bottle of shit vodka. You can make as many lemon drops [for free] as you want.' You have to allow them to be the star, to be the bartender." Brent Puls says that the shrink at the Note was expected to amount to "about 10 percent of the ring." "Everyone always rang everything up even if they comped something. There was a certain amount of trust involved. You'd always comp other bartenders from the neighborhood." At Borderline, this arrangement is implicit, Raul indicates: "The owners know it. They don't want to know it, but they know it."

Such activity "juices the tips" for the server, as local norms require that a free drink be met with an exceptional gratuity, and can in fact be viewed as a cost-effective way for bar owners to improve employee compensation. The reward for the business is further realized in the ability to attract patrons whose mastery of social codes of hipness improves the bar's overall image. Given the markup on drinks, and the fact that a drink given need not imply a drink that would otherwise have been sold, the net cost to the business turns out to be a relatively negligible investment in labor relations and human décor. In enacting this strategy, the owners rely on the discretionary judgment of their employees, which in turn endows servers with a sense of control over the work environment (in contrast with the rigidity of routinized service provision). Still, there are limits to the amount that owners are willing to invest in this outcome, as servers are aware. Says Raul, "You can't give the bar away. The only way they're going to be able to tell is by looking at the register at the end of the night to see how off the register is. If your ring last week was very high and then this week you had a lot of people but the ring was low, you gave the fucking bar away. Then they look at your tip jar, and they see that thing, and they *know* you gave the bar away. [Then] they start watching you closely."

Soft strategies of control allow the server to actively make decisions that translate into production of the bar's ethos. However, if owners perceive that the employee's strategies have become misaligned with the business' goals (extracting surplus value), then control will become more rigid, and if the employee is unable to realign that strategy, he or she gets fired.

The pursuit of tips is a game that servers play, governing a host of strategies of comportment and conferring status as well as income. Within the structure of the work environment, the game enhances intensity of effort, just as Burawoy writes of industrial workers striving to "make out" (meet piece quotas). "The rewards of making out are defined in terms of factors immediately related to the labor process — reduction of fatigue, passing time, relieving boredom, and so on."[16] In the course of a given shift, the game of maximizing tip income serves primarily to foster the sense among workers of being engaged, strategic actors as opposed to just drink-dispensing automatons. Game playing, as Csikszentmihalyi observes, is a social activity especially conducive to producing "flow," in which even mundane, repetitive activities can come to be experienced as highly absorbing.[17] In this way, the grind and multiple indignities of service provision are made bearable.

Bar workers in Wicker Park strategize inveterately to optimize tipped income, a fact that confounds the claims of "just being oneself" through dress and demeanor. Tip pursuit leads employees to tailor their personal aesthetic in accordance with the nocturnal norms of fashion and comportment. Even where this is consistent with an employee's stylistic preferences, it nevertheless transforms self-expression into instrumental action. Krystal Ashe shows how freedom to construct one's own aesthetic amounts to the ability to be strategic in playing the tip game:

I've never worked at anyplace that required me to wear a uniform. . . . I think there's such a great freedom in what you can wear. It's really funny, I was describing to my friend last night, "I'm getting ready for work, I have to wear my tall shoes." She asked why, I said, "I'm working with this tall bartender, and if I'm short nobody's going to come to me, everybody's going to go to her, and she will make all the money." It's weird. You put that much thought into it. I know the reason that I make more money than other people is because I am the only girl working, or on other nights, the only blonde working. I stick out more. If I wear a hat, people come to me because I am the only one wearing a hat. It's just what people see when they walk in the door. They seem to go to whoever's most different. . . . I try to dress the way my crowds want me to dress. I have a really young clientele. Some people would say, "Why are you dressing like that if you're 33?" I don't think I look 33.

Though Krystal began with noting the "great freedom" in what she could wear, she immediately shifted to cataloguing the ways that dress strategies were geared to instrumental concerns, concerns that mirror those of bar owners wishing to sell a hip, creative ambiance. Her comments also alert us to the fact that co-workers, while usually friends, can also be construed as the "opponents" in the tip-optimizing game.

Again and again, informants' sense of "being themselves" in the work environment is complicated by instrumental concerns and social dilemmas. Especially for women, flirting and provocative dress can be a means to enhance their tip potential, but it also can invite attention from male patrons that is unwelcome, while turning off some other patrons. Krystal notes, "A lot of women in the industry make jokes like 'This is my 'paying the rent' shirt.' I don't tend to wear low-cut shirts because I serve just as many female customers who wouldn't find it quite as entertaining." Most of the women working in the local bar scene settle on a strategy in which they try to appear distinctive, without being unduly provocative. Says Amy Teri, "I think when you dress more provocatively, then the women are pissed, and so they don't tip you, and/or the guys are hitting on you, and you're rejecting them, and then they're pissed off about it."

Women in the bar industry must endure exposure to harassment routinely. They find themselves in a tricky situation, since flirting and sexy dress can be effective ploys in the tip-maximizing game but can also lead inebriated male patrons to misrecognize salient boundaries. Says Katie Baker, "Sometimes it definitely borders on harassment. I get a lot of people that grab, people will lean over the bar and touch. I'm like, 'Why are you touching me? You don't know who I am.' I'm tattooed, and people think that's free rein to clutch on my arm and yank it like a chicken bone. . . . For me, I dress up every day, and I'm pretty much a skirt girl. Sometimes that can make it worse, definitely."

At least Katie has the bar as a buffer. Cocktail servers, who work amid the crowds delivering drinks, are almost always women, and they are the most vulnerable to the bad behavior of customers. Amy Teri "worked the floor" before being shifted to the comparative safety behind the bar. She recalls, "Being on the floor sucks royally. You have no protection around you, with guys who want to touch and grab you."

The game of tip optimizing encourages bartenders and wait staff to be attentive to the mundane cues of the service situation. Cocktail servers who work the floor must be aware of where previous drink orders will be drawing low, so that they can replenish the table before the customer takes it upon himself to go to the bar and thus tip the bartender directly. As much as possible, servers will attempt to remember a patron's previous drink order, and may even attempt the gambit of preemptively pouring a "fresh one" before an order is even placed. Moreover, the effective server learns to read the crowd. Says Brett Puls, "You gotta know who needs schmoozing and who doesn't. If you got somebody who's trying to show off to his girlfriend or something, you gotta hook him up, make sure he's getting good drinks. Make 'em all fancy."

Bartenders make it a point to keep careful track of who is tipping well and who isn't, rewarding those customers who are appropriately grateful, and, more important, punishing those who aren't. Shaming is one popular technique. Katie Baker told me, "Sometimes if I have the right confidence and the right amount of time, I'll stand there and stare at them. I finally got the look down. I look out of the corner of my eye — 'put that dollar down.' If I make the right eye contact with them, they will go back into their pocket and pull the money out and put it back onto the bar."

Tactics are often far less subtle, and Amy Teri claims that she will occasionally yell at deadbeat customers. Especially when the bar is busy, bartenders have considerable discretion to withhold service from certain patrons. Says Katie, "Tipping is necessary. It is. I will tell people if they're being cheap. I will not serve people if they're not tipping. I will tell other bartenders [not to serve them], too. If multiple rounds have gone by, and [the customer] hasn't left me anything. I'm not going to waste my time on [him or her]. I have other people who will give me their money, and I will go to them, priority. I'm there to make money. I'm there to make more than $5.50 an hour. That's what I'm there to do." Amy likewise shows little forbearance to patrons who fail to reward her efforts. "Quite frankly, if you come in here and you don't tip me, you're not getting waited on. I will full-on tell you, 'Money talks, and bullshit walks, and if you don't have money in your hand, you can leave and I don't care.' This is how I make a living, and everybody knows this. We're not in China. TIPS: To Insure Proper Service." As Amy's comment indicates, patrons with any nocturnal savvy should understand how the game is played and hold up their end. As far as she is concerned, if they fail to do so, they have earned shabby treatment, and aren't the kind of patrons the bar wants anyway. Quite possibly she is right. Though owners would be unlikely to actively encourage behavior that leads patrons to take their money elsewhere, the fact remains that the hip reputation of the place relies on a crowd that is for the most part well versed in bar rules, that is, that possesses hip cultural capital. This sort of patron knows to tip well (none better than other service workers). In any event, while most bartenders I spoke with confirmed that they sometimes employ these aggressive tactics, none reported ever being reprimanded by their employers for doing so.

Thus, withholding as a mode of informal social control in Wicker Park bars goes on with the tacit approval of bar owners and managers, so long as the bartender keeps the ring at standard levels. Allowing the bartender this shallow social authority helps maintain "soft" control over the workforce. Bartenders control the terms of their work environment to a necessarily limited and contingent extent, in the process playing a game whose score is registered in the proliferation of dollars in the tip jar. The reward to owners comes from intensified labor. Says Raul, "It's easy to forget your job, because it's such a fun place to be at." But the absorption, if not the fun, that comes with playing the game blocks out such distractions, reminding bartenders that, as Raul explains, "you have to do your job. The minute the bartenders are not doing their job is the minute they're not making money. And the bartenders are the workhorses, they're the front lines."

Why Bartenders Make Less Money Than They Think

The game of tip optimization thus helps to absorb service workers intensely in their menial tasks. Does it also provide a reasonable income? Bartenders that I spoke with report that in fact a person can make a pretty good buck in the Wicker Park bars. Says Krystal Ashe, "Artists and writers work in this industry because where else can you work for four hours and make $200, $300, even $400? If you're a waitress [as opposed to bartender] you can make $100 or $200 in four hours. That's why artists flock to the service industry. You can support yourself on twenty hours a week." These estimates, which vary from job to job and from night to night, indicate that the compensation for service labor, coming almost entirely from tips, far exceeds the averages expressed in official service sector statistics.[18] Other service workers in the neighborhood estimate incomes that fall within the broad parameters Krystal stakes out. The money looks even better given that most industry workers interviewed own up to the fact that "Uncle Sam" is generally cut out of most of it — that is, service workers neglect to report their incomes in full for tax purposes. As one bartender put it, "I can tell you what I make in a year. I probably make between — I've been fluctuating back and forth between bars — I can make $40,000 a year bartending, working four days a week. Only half of that is pretty much declared, that's nice."

Tips therefore allow middle-class youth to earn incomes far in excess of their cost to the employer, which typically amounts to minimum wage and some comped drinks. However, extended observation and interviews indicate that workers in these jobs are imperfect accountants when it comes to estimating their own incomes. The uncertainty about actual income is apparent in the quote above, in which the informant shifts quickly from a promise to reveal what she does make to what she probably makes and finally to what she can make. Income estimates typically are distorted by the perhaps willful tendency to ignore the substantial contingency involved in tipped remuneration. The rates of tipping can vary enormously depending on the time of evening the person is working, making some shifts much more profitable than others. Different days of the week also generate wide variations in income. Part of the service job involves setting up for later or cleaning up at the end of a shift, tasks that do not earn tips. Moreover, many service workers do not have fixed schedules, and may get to work plum hours only sporadically. Especially for establishments that have outdoor seating, there is significant seasonal fluctuation in labor requirements. Vacations are unpaid. Even those who are permanently assigned desirable nights like Friday and Saturday typically must also turn in shifts on much less remunerative nights.

But despite this variation, service workers often speak as if their best hours are actually their average hours.[19] This could lead one to extrapolate weekly and annual incomes that are much higher than those actually realized by employees in Wicker Park's bar and restaurant industries. Further, workers in the industry often achieve their peak in earning potential fairly quickly, after which earnings plateau or even decline. Thus, even where bartenders do earn more money than entry-level professionals, in five years the bartender's income will be roughly the same, but the professional's will have improved substantially.

Moreover, earnings from tips are further complicated by the norms governing the behavior of bartenders and other servers when they themselves go out. Service workers in Wicker Park are inveterate consumers of the local nightlife on their off nights, frequenting establishments where they are well acquainted with the staff. On these forays, they are the recipients of "professional courtesy" in the form of free drinks. However, this does not serve to defray the expense of inordinate participation in the entertainment economy in any meaningful way, since free drinks are met with tips that generally amount to more than the price of the comped item. Service workers tip better than ordinary people do — much better. In Wicker Park, service workers frequent the bars, and tip income is inflated by the extraordinary gratuities they leave. Since the norms for participants in the industry require that this largess be returned in kind, that portion of their income contributed by other people in the industry is devoted to a ritual exchange of money, the true meaning of which is to be found in the performance of status among participants.

The tipping behavior of local service workers thus turns out to be a circular process of mostly symbolic exchange. Says Krystal, "I work service industry nights, so I get people tipping $10, $20 a round. [But] believe me, I'm in the same boat. If I go out, I have to tip $20 a round. If you're going to get one drink, and you're a bartender who just works a so-so bar, you're probably going to tip me $2 to $3. If you are a bartender who works at a really great place, you're probably going to tip $5 to $10 for one drink. That's how it works." As Krystal indicates, both tipping and the attendant "comps" are a central part of the performance of local status. Bartenders at elite spots, the ones that are most popular and respected, expect and receive a larger amount of preferential treatment. But they are also supposed to "represent" by laying out outsize gratuities. The perks of "free drinks," given as a courtesy to those known to work within a given circuit of bars, turn out not to dramatically alleviate the financial burdens of overactive party lives. According to Matt Gans, service industry workers from the elite spots "pay for less. You'll get five drinks for free. But the money you would have paid for drinks, you'd better tip. You may get drinks, but you're tipping forty bucks a round."

The more a worker is known and numbered among the elite in the neighborhood nightlife, the greater the onus to give and receive monster tips. Thus, when it is an interaction between status equals, the exchange resembles the wagering patterns that Geertz describes in the Balinese cockfight: "So far as money is concerned, the explicitly expressed attitude towards it is that it is a secondary matter. . . . [The wagerers] mainly look at the monetary aspects of the cockfight as self-balancing, a matter of just moving money around, circulating among a fairly well-defined group of serious cockfighters."[20] Like the cockfight, then, what the tipping game "talks most forcibly about is status relationships."[21]

Amy Teri indicates that her tipping behavior, while always exorbitant by any reasonable standard, varied heavily depending on whether or not the person serving her also came into the Borderline when she was working and thus directly participates in the symbolic exchange that governs local tipping behavior. "I tip a lot to people who come into my bar. If they don't come into my bar, then I tip average. Two bucks, five bucks, ten bucks. If there is somebody that comes in all the time, I tip 40 bucks." Amy acknowledges that this behavior has little rational justification: "I don't know why industry people tip each other. It's stupid. It's like a big pissing contest over who can tip the most. It's really stupid, and it's especially stupid when bartenders who work here tip the other bartenders who work here. It really is a big pissing contest." Puls likewise notes the circularity of the tip game: "It's really funny, all the bartenders in the city make like two hundred and fifty bucks [a shift], but if they go out, they have to spend a hundred bucks on tips to the same bartenders. So it goes right back into the other bartenders' pockets. You know, there's all this circular motion, no money is made, just changing hands."

The norms of circulation that attach to these tips thus suggest that they should be significantly discounted, since a large amount of the excess will simply be returned into a more or less closed local economy of tip exchange. Those who give in to the social pressures to make the scene on off nights — and the overwhelming majority of service workers that I observed do so — thus find themselves spending their incomes (helpfully received in cash) as fast as they collect them. This is how so many local service workers can claim such impressive incomes (though not on their taxes) and yet still find themselves pinched at the end of the week for things like art supplies, groceries, or rent.

La Vie de Bar

We can see that for neo-bohemian service workers, leisure activity is patterned by their conditions of employment. Their identities both as artists and as workers in a service industry provide a vested interest in differentiating themselves from normal customers and particularly the young professionals they wait on. Unburdened by 9-to-5 jobs and usually working on Friday and Saturday night, employees in the bar and restaurant industry prefer to go out on weeknights, when they will be less infringed upon by regular people, who they refer to derogatorily as "amateurs." A circuit of destinations evolves, with different bars favored on Sunday through Thursday nights. These are called "industry nights," and they have a certain cachet that also attracts other patrons who pride themselves on being "in the know."

Service workers, frustrated by being on the wrong end of nighttime revelry, often find ways to go out on the nights that they work as well as on off-nights. Shifts vary across different venues and types of service sector employment. Restaurant workers typically complete their duties by midnight or 1 A.M., with several members of the wait staff being "cut" even earlier, following the dinner rush. Different bars have different closing times approved by the city. Many are legally required to close at 2 A.M., while others are allowed to remain open until 4 A.M. Sunday through Friday, and 5 A.M. on Saturday. Bars with the later closing times become popular destinations for service workers who have completed shifts earlier at other venues. During the week, the Borderline, which is licensed as a 4 A.M. bar, will typically attract patronage from the staffs of local restaurants and bars that closed earlier. Betty's Blue Star, a Ukrainian Village corner bar on Ashland and Grand Avenues, is also popular with the workers in other Wicker Park bars, and 2 A.M. typically signals the beginning of a rush in which both ordinary patrons and employees of establishments that close earlier migrate in.

Though it is ordinary at many bars for the employees to drink while working, they generally do not get excessively inebriated on the job. But in the so-called after hours bars they strive to "get their drunk on," feverishly making up for lost time. There are also a small number of clubs that flout the city's mandated closing time, particularly an establishment in the Clybourn corridor known as the "Cop Bar" because members of the city's law enforcement community patronize it heavily as they wind down from their own graveyard shifts. Loft and house parties planned for after hours are also popular with bar and restaurant employees. Advertised by word of mouth, these are more exclusive, allowing workers to avoid rubbing elbows with those that they may have served earlier in the evening. This is an extra advantage on Fridays and Saturdays, the nights that service workers consider bar patrons to be especially boorish and socially unattractive.

Anne appeared sheepish when asked how many nights a week she went out. "Three or four nights a week," she said, then amended, "I go out three." Given that she also worked until closing on Friday and Saturday nights, this signals an extremely grueling nocturnal regimen, one with obvious health consequences. Anne continued, "Now that I'm working [at Borderline], I'm going out more, I'm smoking more, I'm drinking more. It sucks you in, and it's hard to stay away from it all. It's like this big social thing, Everybody talks about everybody else. It's hard to stop socializing once you get into that sort of scene, the bar scene." As this indicates, the desire to patronize local bars is abetted by the fact that working in the industry generates a network of relationships that extends beyond one's specific place of employment. Turnover is high, and Wicker Park bartenders and wait staff typically have many former co-workers now employed in other local venues. Gans, who managed the West Side establishments Big Wig and Mirai in Wicker Park and the Funky Buddha Lounge in neighboring River West, recalls, "When I was [in the industry] we went to every single club that was around. We'd bounce around. We have a friend named Harold, so we gotta see Harold. Wherever Harold was. Or Hugo is bartending, or Richie. You go where you're going to get in the door fast. Get in, get your drunk on. Returning favors. [You say] 'Here's where I work.' People come into your place, you go to theirs." When going out, neighborhood workers in the bar and restaurant industry are under pressure to perform competently and separate themselves from the amateurs. Status in the local scene is registered by recognition at other venues that one is an insider and thus entitled to special considerations. These considerations bring with them a heavy obligation to evince appropriate comportment — that is, in terms that are nebulously defined yet clearly recognized, to be cool. Since other service workers should be aware of the pressures that bartenders are under and of the hard work that they perform, they are expected to be able to order efficiently and display appropriate gratitude. Says Amy Teri:

It amazes when people that I know for a fact are in the industry can't even order a drink. Normally, there's etiquette of bartending, and people order their beers all together, and their drinks all together [efficiently]. But you get people who order however which way and say "um" between each sentence and have no idea what you're talking about, and you have to think, "You [the customer] are really dumb. What bar do you work at?" People who come in that are in the industry, if they don't have their money in their hand, they will get charged extra.

Failure to comply with expectations results in the loss of ordinary favors, known as "professional courtesy" and in loss of local regard. Even the places of employ lose status when their workers behave badly on their off nights, for failing to have properly acculturated them to the norms of the night.

Brent Puls, whom no one would accuse of being anything but a smooth operator, demonstrates the costs and benefits that go with participating in the off-hour party circuit:

> When I go out to bars, I pretty much only go to places where I know people, and I'll drink for free. Because there's a network of people that know people. Especially being at the Note, the reason that you hooked up everybody at Pontiac [Bar] is because people at the Pontiac knew who you were, so of course. . . . You go out quite a bit in the industry. I keep telling myself not to. But it's kind of nice in the neighborhood, because any time I'm really bored I can get off my ass and go to the Note, without calling anybody or making plans with anybody and know there will be some people there I can talk to. But yeah, I probably go out three nights a week [in addition to the nights that he works]. And when you're out, you have to represent big time.

Some explicitly link the demands of going out to their job, involving the instrumental desire to maintain a high profile in the nocturnal community. When out socializing, bartenders and wait staff may encourage others to come to their establishment on nights when they are working, helping to maintain the popularity of their venue through this informal advertising, while hinting at the possibility of favor exchanges. Says Krystal Ashe, "It's part of the industry. You go out to promote your nights, and you go out because all of your friends want to go out." Here partying appears as an extension of work. As Neff notes of fashion models, another occupation in which an active nightlife is apparently de rigueur, "Networking means that playtime is no longer a release from worktime; it is a required supplement to worktime, and relies on constant self-promotion."[22] Krystal perceives that this double duty increases her income potential, but she does not factor these informal ventures into her accounting of total weekly work hours, and she does not discount her weekly income by the substantial amounts that she spends spreading tips around while "promoting." The real winners are the bar owners, who benefit from Krystal's high-profile persona, whether she is behind the bar or improving the standard of the patronage.

That Dark Night of the Soul

The bars in Wicker Park are places where local artists and intellectuals engage in debates about art and ideas, debates that in my observation tend toward increasing incoherence as the evening progresses. These lofty exchanges compete with other priorities, such as the pursuit of sexual liaisons and other sensual pleasures. Such objectives do not differentiate the hip, arty set from the yuppie patronage that crowded the local bars as the 1990s wore on, though local service workers are unanimous in their view that scene insiders are better behaved. Nevertheless, artists are hardly immune from making decisions under the influence of drugs and alcohol that they later regret. Indeed, Liz Phair comments ruefully on the emptiness of the neighborhood's sporadic sexual encounters in the song "Fuck and Run" from *Exile in Guyville*. Wicker Park hosts a nightly Bacchanalian festival of often reckless hedonism, with important consequences for those who make it the scene of their labor. The opportunity to make money while inhabiting a superhip nighttime environment is a major attraction for the young artist/service worker, but it is also the source of serious occupational hazards.

The sexual charge of the bar scene, lubricated by alcohol and other intoxicants, generates both welcome and unwanted sexual attention for workers. Female employees in particular complain about grossly inappropriate behavior by aggressive male patrons. Male bartenders express far more tolerant dispositions toward their flirtatious female customers. Brent indicates, "On the weekend sometimes I'd wear a nice shirt, so that maybe a hot girl would come up and hit on me. Which happened maybe 50 percent of the time. And also it would kind of juice the tips." In general, the opportunity to meet potential sexual partners is viewed by men in the industry as more of a perk than a drawback. Women also flirt while working, in part for instrumental tip considerations, sometimes just to pass the time. Says Amy:

> Everybody flirts. As far as bartenders going home with people at the end of the night, at this bar not so much. At other bars, definitely yes. Some people don't even wait to leave the bar. The bathroom has been frequented. Not by me. I want to say that for the record: not by myself. But the owners, the bartenders, the staff of any kind. People who don't work here at all. The women's bathroom, have you ever been in there? It's not like the men's. The women have these steel dungeon doors, and it must turn everybody on because everybody has sex in those.

Sex is not the only illicit activity that takes place in the Borderline bathroom. Like many bars in the neighborhood, Borderline has a well-known reputation for drug use: "It's always been known as the Snort-a-Line," Katie Baker told me, referring to the heavy cocaine intake among customers. Before bartending, Raul worked security at Borderline, and he indicated that efforts to curtail drug use were fairly halfhearted: "There's a lot of coke floating around. It's ridiculous. You try to catch them, and it's like: 'Man put that shit away, go have fun.' But people have vices, man, that's it. You can't hate anybody for having vices. They're going to do what they're going to do. Once I walked in on a guy doing smack [heroin]. We threw him out. That's classless. It's a little too much of a procedure." On several occasions I also witnessed furtive, and sometimes not so furtive, drug taking in the bathrooms of the Borderline and other bars. On one occasion, I saw a couple of young males get caught by a Borderline doorman (not Raul) sharing hits from a vial of cocaine. The doorman asked for some, and then told them to wrap it up after his request was satisfied.

Working in the bar industry exposes employees to inordinate opportunities to take drugs. Working while high on marijuana is so common that it is taken for granted — almost everyone I spoke with acknowledged that they had co-workers who imbibed before a shift or even on breaks, and none felt that this was a major problem. Cocaine use was taken more seriously, but informants concurred that it was also fairly common, if not specifically in their place of employment, then certainly industrywide. Says Krystal Ashe, "Coke was out for so long, and then coke came back in. There are some bars in the city where the whole staff is going to be yacked up, and there's other bars in the city where the management goes on Ecstasy and GHB (a club drug) just to take it [deal with the pressure]." Amy Teri denied that there was a major drug problem with the Borderline staff: "I think everyone here is pretty clean and sober. Even here you're smoking a little pot or whatever, but other than that there are no major drug problems in this bar, which is a rarity. Other bars have major problems. I've worked at a few other bars, and cocaine is the number one enemy of the majority of bartenders."

In fact, it appears likely that alcohol is the number one enemy of the majority of bartenders, both in Wicker Park and nationwide. In Wicker Park, heavy alcohol intake among service workers is so common as to be taken for granted, though in the context of national data on adult alcohol consumption its eccentiricty becomes clear. Data from the 1997–1998 National Health Interview Survey (NHIS) indicates that only about a third of adults ages 18–45 consumed five or more drinks in one day at least once in the previous year. For women the rate was less than one fifth of the population between 18 and 45.[23] Conversely, the bartenders and wait staff of both genders that I observed in Wicker Park typically consumed five or more drinks most nights of the week, virtually without exception. They seldom acknowledge that this behavior might be a little bit problematic.

In 1997, the National Household Survey on Drug Abuse (NHSDA) indicated that food preparation workers, waiters, waitresses, and bartenders collectively had the highest rates of any occupational category for both current illicit drug use and heavy alcohol use, at 18.7 percent and 15 percent respectively. Rates among industry workers in Wicker Park appear much higher. At most local bars, blatantly open drinking on the job is tolerated by owners and managers so long as it does not clearly diminish performance — among the establishments that I observed most closely, only Mirai, owned by the unusually authoritarian Meia, was an exception. Indeed, drinking is often a standard part of the scene-making interaction with customers. Doing a shot of hard alcohol with a customer is an ordinary, accepted practice. Says Teri, "If somebody says, 'Let me buy you a shot, you will do a shot.' But as far as getting excessively wasted, no." Likewise, Puls indicates, "At the Note everybody drank on the job. Everybody would do shots with the customers."

It is clear that these workers have much higher than ordinary tolerance for alcohol, and it would take a lot for them to feel "excessively wasted." This becomes part of the status identity of industry workers, by which they distinguish themselves from ordinary patrons. Amateurs are those who behave sloppily when drinking or doing drugs, while industry professionals pride themselves on remaining competent even after heavy intake. Workers do not view their performance as impaired even after they have consumed amounts that would put their blood alcohol ratio well over the legal limit for, say, handling a motor vehicle. They imagine that others cannot even tell that they are drinking or drugging. Still, Katie Baker indicates that "there are a couple of employees who drink way too much. At work, it's really aggravating because you can see the effect. It slows them down. Their drawers are all screwed up, they get clumsy, they get in the way."

The norms of bohemia also contribute to the vulnerability of these workers to excessive drug and alcohol consumption, since, as we have seen, substance abuse is less likely to be negatively sanctioned than in other milieux, at least if the user can maintain a certain level of performative competence while under the influence. The temptations to participate in these activities are compounded for artists who take jobs as service workers, and are thus surrounded by other revelers. Older, more experienced service workers indicate that young, green employees present the biggest problems. This appears to be because these workers have not learned to incorporate substance abuse into their habitus such that they can still be functional while working. Says Krystal, "Those are the people that get the most caught up quickly and make stupid moves. I'd rather hire a manager that was an alcoholic than someone that was new and didn't drink and didn't do any drugs because that person is going to be most susceptible probably."

Everyone has a story of some co-worker that became a casualty of the temptations that come with the scene. Raul explains:

I've seen people go downhill, get into dire straits, because they took it for more than a job; they let it get into their personal lives. They took it personally. You see people getting fucked up, you're going to get fucked up. Some people at work, while they're working they're drinking. Easily done, but it's an easy way to lose hold of yourself. It becomes a habit like anything else. It's all habits in there. It's all people's vices exercising themselves. People get rid of what they want to get rid of, emotionally, spiritually. Shoot their load off, get the fuck out, you know, drink and get fucked up, and forget who you are for another night. And working in a place like that can drain you. It makes you very depressed, and you pick up habits of your own. And sometimes you lose your cool and lose grip of who you are.

For most participants, bars and nightclubs are liminal spaces[24] where they can throw off ordinary inhibitions, often recklessly pursuing intoxication and sexual objects. However, this carnivalesque atmosphere poses challenges for those who make it the scene of their labor. In this context, boundaries are hard to maintain. Industry workers are on display in spectacular leisure-time apparel, grooving to music that they themselves may have selected, perhaps angling for sexual partners, perhaps participating in the intake of drugs and alcohol. In these ways, their labor may bear a striking resemblance to play. But if this is the case, it is a game that can have exceptionally high stakes. It is what Geertz calls "deep play": "play in which the stakes are so high that it is, from a utilitarian standpoint, irrational to engage in it at all."[25]

But people do play, despite the drawbacks of foreshortened careers, limited upward mobility, overexposure to drugs and alcohol, and, as time goes by, a general feeling of exhaustion and malaise. The attraction of the game comes from its apparent compatibility with the norms of the bohemian ethic: the exaltation of the sensual and the aesthetic, the resistance to "corporate" styles of labor, commitment to flexibility, and the desire to be in the scene, whatever the terms. It's a potent elixir, the gritty neo-bohemian glamour of the neighborhood nightlife, but one that can also produce a nasty hangover, as Raul reports:

The bar culture is the bar culture, you know. I've been pulled into the bar culture, and I mess with it, but it's a tough business to be in. It could be a little demanding at times, because when you don't want to talk to people, sometimes you have to. And it can drain you. Now I'm trying to get back to things I did with a passion. You can't really drink your life away, smoke your life away, and I'd forgotten about all my ambitions. I think a lot of people fall into that. It's like trying to go to sleep, and you can't go to sleep.

CHAPTER 9

The Digital Bohemia

Making money is art, and working is art,
and good business is the best art.
    —Andy Warhol, *The Philosophy of Andy Warhol*, 1975

New Media in the New Bohemia

THE OFFICE WINDOWS ON THE SIXTH FLOOR of the Northwest
("Coyote") Tower provide a panoramic view of the Wicker
Park neighborhood and — in the distance, straight down
Milwaukee Avenue — of the looming skyscrapers of the Loop.
Periodically, one hears the pleasant rumbling of the El Train
pulling into and out of the stop a block away, a still vital "retro"
emblem of Chicago's place identity tethering the neighborhood and its various manifestations of hip commerce to the
Loop's glistening fortresses of capital and, in the other direction, to O'Hare International Airport. Entering the 1990s, the
Tower's offices lay largely dormant, but as the decade came to
its close they housed numerous ventures in the neighborhood's
proliferating media design sector. It is a familiar paradox, a
retro space housing the high-tech commerce that sprouted in
Wicker Park at millennium's end.

It's January of 2001, and I'm being given a tour by Michael
Weinberg, a self-proclaimed "retro-futurist" and founder of the
Internet firm Buzzbait, which occupies the Coyote's sixth floor.
"It's not a very high-tech building," he says of its decaying
amenities. Likewise, several of the old industrial lofts scattered

Figure 9.1 The Northwest (Coyote) Tower. Photo by the author.

throughout the neighborhood are now home to design firms, many of
which exploit the cutting-edge technologies of the informational economy
in their contribution to the contemporary deluge of signs. Buzzbait is a boutique firm specializing in interactive web design for a small but impressive
roster of corporate clients, including the local Tribune Corporation.[1] This

same view graced the cover of that month's issue of the *Industry Standard*, an Internet trade publication, which announced that the neighborhood has become "the best new place for media companies"[2] like this one.

Wicker Park would have seemed a poor candidate a decade ago for the current proliferation of high-tech enterprises. But, as we have seen, in 1989 the Northwest Tower lent its nickname to the "Around the Coyote" arts festival, advertising the young artists who lived and worked in the neighborhood, and attracting still more. The artistic presence was crucial in attracting digital enterprises as the decade reached its close. The neighborhood became associated with "design distinctiveness and fashionability"[3] in marketing practices facilitated by the Internet. A 2001 cover story in *E*Prairie,* an online publication devoted to Midwest technology news, announced that "tech artists find Wicker Park . . . great for business," adding that "what a decade ago was hailed as [the] Chicago equivalent of New York City's SoHo District has now become digital."[4] Was the new bohemia in Chicago the new Silicon Alley as well?

The boosterism evinced in the *Industry Standard* and *E*Prairie* accounts of Wicker Park's design sector is standard issue for a genre of business reportage whose default mode has always been hyperbole, with everything inevitably the best and the newest: otherwise, why would we care? But this hype turned out to be a feeble afterimage of the spectacularly overheated mania that accompanied the tech boom of the 1990s. Indeed, the *Industry Standard* itself was soon to become a casualty of the Internet stock crash that wiped out so many of the companies that the *Standard* had celebrated in its four years of existence. While dot-com advertising brought in $140 million for the magazine in 2000, allowing it to turn a profit, this exorbitant figure fell to earth the next year with the same dizzying speed as the NASDAQ. In 2001, advertising revenue declined by 75 percent, and on August 8 the *Industry Standard* published its last issue, only six months and change after it broke the "news" on Wicker Park.[5] Recent history has been kinder to the neighborhood tech scene than to its media cheerleader. Few of the new firms in the neighborhood had been brought public, and so the stock market crash had little effect on the sector. Wicker Park looks different from previous "best new places" that rode the boom — San Francisco's South of Market (SOMA) district, for

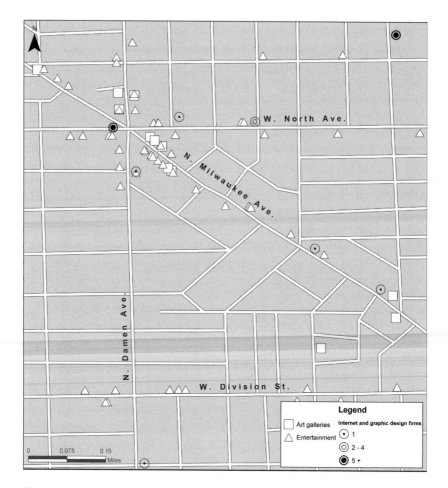

Figure 9.2 Art, Entertainment, and Design in Wicker Park, 2001. Map created by Richard Lloyd and Todd Schuble.

example — and these differences may alert us to the real future of work in the world of digital design.

Chicago's near West Side was home to a significant number of media firms that included Internet content among their services; I was able to locate two dozen such firms in 2002 operating in and around Wicker Park. These enterprises were interspersed with the art galleries and entertainment venues we have already examined in the new bohemian milieu (figure 9.2). New media enterprises are attracted to the neighborhood for both highly tangible and more abstract advantages, as the *Standard* article suggests:

"Compared to downtown, commercial rents are cheap. The area abounds in graphics firms for subcontracting jobs, and the location serves as a kind of geographical shorthand that conveys nonconformity and innovation."[6]

But being the best place for media companies — or simply being a good place anyway — means something quite different in 2002 than it had in the preceding few years. Wicker Park's media design presence did not attract the flow of venture capital that once visited Silicon Valley or, to a lesser extent, New York's Silicon Alley.[7] For that, ordinary residents of the neighborhood should probably be grateful, given that many claim the astonishing rise in rents that came with Internet commerce in the Bay Area ruined San Francisco.[8] The Wicker Park firms I examined were privately held, and combined Internet expertise with more mundane graphic design and marketing services. During a conversation in 2000, Weinberg expressed his admiration for New York's cutting-edge design boutique Razorfish, which had only recently seen its stock valuation soar to more than $4 billion after going public, and it was clear he harbored aspirations to follow this path. But by 2001 Razorfish's share price had dropped to less than $1.[9] At Buzzbait's one-year anniversary party in 2000 — this was my first visit to their offices — speculation about buyouts and IPOs animated conversation. A year later, that had been pretty much quelled.

Still it is worth asking how a formerly industrial Chicago neighborhood became the home to a high-tech sector at all, albeit one at a more modest scale. After all, conventional wisdom in urban studies argued through most of the 1990s that digital enterprises locate in Sunbelt cities and follow spatial dynamics different from those of the industrial production that fueled Chicago's growth. We have seen that older center-city neighborhoods in cities like Chicago are often viewed as left over from another period of capitalism, destined either to be mired in self-reproducing cycles of poverty or gentrified into culturally homogeneous, elite residential and consumption enclaves for professionals — that is, as anything but production sites. Moreover, information technology is commonly identified as accelerating the irrelevance of older spatial arrangements.[10] Why, then, would boutique Internet enterprises find Wicker Park to be a strategic locale of operation?

Once again it comes back to the surprising role neo-bohemia plays in organizing creative labor. Scholarship on the technology-driven economy

Figure 9.3 Buzzbait's Holiday Card. Courtesy of Michael Weinberg.

emphasizes the importance of "talent" to the success of firms, that is, of the labor of individuals referred to as "knowledge workers"[11] or "symbolic analysts"[12] who possess the creative competence required by technology and design enterprises. Many argue that access to such labor has become the key to understanding the location choices of firms, replacing the older spatial variables of fixed capital and transportation that characterized industrial production. These studies emphasize the importance of local amenities as a magnet for highly mobile knowledge workers, employing a number of methods and measures to operationalize "quality of life."[13]

Using a statistical measure he calls the "bohemian index," Richard Florida finds a robust correlation between the presence of artists in a region and the concentration of high-technology enterprises.[14] "The . . . presence and concentration of bohemians in an area creates an environment or a milieu that attracts other types of talented or high human capital individuals. The presence of such high human capital individuals in a region in turn attracts and generates innovative technology based industries."[15] Following this logic, Wicker Park's neo-bohemian cultural profile would be viewed as resource for firms insofar as their workers value it as a consumption amenity. However, examination of new technology developments through the heuristic of neo-bohemia narrows the focus from region to neighborhood and moves beyond the emphasis on local culture as a simple amenity.

The small firms in media design that concentrate in Wicker Park generally provide services to larger corporate enterprises on piece assignment, producing contingent and uneven revenue streams. Dealing with such uncertainty requires flexible access to creative labor. The small firms in the expanding media design sector use local artists as flexible labor and draw upon the local ambiance of creative energy. Many workers in the local design sector have formal training and/or are active participants in fine arts, music, or theater. They minimize the differences between media design on behalf of corporate clients and other artistic pursuits, seeking to construct their design work as continuous with an ethic of bohemianism, even as that construction is challenged by the fact that their labor is ultimately exploited for profit by capitalist enterprises with which they do not identify.

Similar developments are evident in neighborhoods like Manhattan's East Village or San Francisco's South of Market, where media enterprises translate a local ambiance of hipness, the urban imaginary of "bohemian cool," into images available for global diffusion. However, as economic development intensifies in such neighborhoods, contradictions arise in this production process. The homogenizing tendencies of gentrification threaten to undermine the local distinctiveness that is exploited as a source of value for media enterprises. With the influx of venture capital that characterized the late 1990s effectively choked off by market reversals, increasing ground rents make some neighborhood areas untenable for small innovative enterprises. As Joel Kotkin notes, "High real estate prices and tight labor markets are making some of the most successful urban neighborhoods too expensive for growth companies."[16] Like the new bohemia, the new media district represents an apparently fragile convergence.

Neo-bohemian culture is a factor of production for new media enterprises, inflecting the creative work of local designers. The "economies of signs and space" described by Scott Lash and John Urry consist of more than deracinated informational flows;[17] they entail material relations of production decipherable through distinct sites of social activity.[18] Thus, districts like Wicker Park are organized in the service of aesthetic production, including both traditional arts production and activities in the new media. These aspects of the spatial field interact with one another, linked through the downtown business district into global chains of image circulation.

From Subculture to Global Culture

I sit uncomfortably eyeing the open space in the first-floor offices of Boom Cubed, a Wicker Park design and marketing firm on North Avenue, listening to the drone of techno music coming from oversized speakers. My perch is in the makeshift reception area, which is just a corner tucked to the side of the many computer terminals where the firm's employees toil. I sink into the couch; around me are blue plastic chairs shaped like hands, resistant to sitting but consistent with the determinedly retro–future shock ambiance. Periodically the phone rings throughout the office, and whoever answers yells out a name to complete the connection. Often the employees — six are active as I sit there — cluster around a terminal, commenting on a work in process: "that's radical," "that jams," "that's dope," or perhaps offering creative input. Then they zip back to their own projects. I write things down, trying to be invisible. With embarrassing frequency someone asks if I need anything. They are clearly unsure what to make of a sociologist in their midst.

It is the early spring of 2001, and Boom Cubed has recently merged with a marketing enterprise called Phil and Nate, a small firm of young African Americans that sells marketing services with an "urban" aesthetic. The merger enhances the already impressive racial and ethnic diversity of the firm. Plans to change its name to 01-11-1 (Oh-One-Eleven-One) — after a line of computer code — are in the offing. The workers around me are all young — under thirty — and dressed in the hip-casual attire that has become the signature of the new economy workplace. By appearance they could easily be taken as young artists working the neo-bohemian beat in the neighborhood, and indeed such a guess would not exactly be inaccurate. One of the employees doubles as a performer in a local punk rock band, and another one deejays on Tuesdays around the corner at the Artful Dodger, a weekly event that typically serves as a company night out.

Begun in 1995 by two young design graduates from the University of Illinois, Boom Cubed went from a shoestring two-person operation making fliers for musicians in the "underground" arts scene to a firm employing two dozen workers and contracting services to such global behemoths as Nike. The development of Boom Cubed is instructive for understanding the emergence of the design presence in Wicker Park, illuminating its linkages both to the local arts scene and to the downtown advertising sector. Boom Cubed specializes in the provision of hip, multicultural content, aimed at the youth market with which the firm's young, multiethnic employees presumably identify. Brad Cowley, the firm's co-founder, put it this way:

> We know that age group. We have so many different people here with so many different [cultural] interests. We use that knowledge, and we encourage that knowledge to be brought in, and brought back to the client, to say, "Look, this is what our understanding is of the audience that you are trying to talk to." In that context, it's a great asset, because it really does help sell our message, what we are trying to do.

This understanding of the youth market reflects a particular experience of urban youth culture grounded largely in the neo-bohemian spatial practices of the neighborhood.

The seeds of the enterprise were sown in informal collaboration between Cowley and Jennifer Arterbury immediately after their graduation from the Graphic Design Program at the University of Illinois in the early 1990s. Both supported personal creative interests in their immediate postgraduate years by working with advertising firms located downtown. Arterbury explains:

There's a lot of work for young designers. . . . I mean you've got huge agencies, they're always hiring designers. People move from place to place, job to job. I freelanced for an agency called, they changed names, it's now T&P Worldwide, which is a recruitment ad agency, and they bought a whole bunch of smaller agencies all over. But I had an internship at J. Walter Thompson [in college] with a woman named Christina Oravan, and she was sort of my mentor, and when I moved here and started looking for some freelance work . . . I started sending out resumes, and she hired me to come out and work right away. So I did, and that continued as freelancing, and then I took an art director's position there, and that was it.

Though finding work was easy, Cowley and Arterbury were dissatisfied with the creative limits of their day jobs as designers. Their common interests in Chicago's youth culture and arts community led them to seek after-work outlets for expression.

Arterbury notes the importance of the neighborhood to developing these interests:

[Wicker Park] is a great area, there's always stuff going on. . . . Well, first I moved down here to be closer to work initially; [that] was when I moved [to] off Chicago Avenue [in Ukrainian Village], and then I moved here [to Wicker Park], because initially I used to hang out here a lot anyways. We would work on a couple magazines, and we would always meet over here, and we did a lot of hip-hop work, and we did a lot of publications and fliers for that kind of thing. There are always parties over here [in Wicker Park] for that.

Participation in the "underground" neighborhood scene, rather than profit potential, initially motivated the extracurricular activities of these young designers, activities that provided the foundation for their boutique shop that now serves the corporate "mainstream." Cowley recalls:

When we first started it was more underground. A lot of underground music. We were going to clubs all the time, doing fliers for shows. It was underground, it was hip-hop, it was connections that we knew, it was music that we liked, [and] it was shows that we were already going to. It ranged from big shows at Metro with the Fugees, Diggable Planets, Queen Latifah, stuff like that, to smaller shows with Chicago artists at places like the Bop Shop. We weren't making a lot of money, but it was a good time.

As the collaboration expanded into more ambitious territory, with offices and employees, low-paying jobs on behalf of the local arts community were largely phased out, Cowley said.

[Hip-hop fliers] pay, but they don't pay a lot. Now, we don't like doing fliers anymore, because they don't pay. If it was just Jen and I, it would be okay. We would find a way to survive. But now we've got fifteen, twenty people working for us, we've got to make sure we can pay their salaries. Pay insurance on the office, pay benefits. It all has to be answered. And we don't answer that by taking jobs that don't pay. So by necessity we've shifted the focus of our company.

Although such jobs would become undesirable as the collaboration expanded from hobby to business enterprise, the "for fun" work doing hip-hop fliers was a crucial factor allowing Brad and Jennifer to develop skills that would prove valuable to larger, better-paying clients. Increasingly, Jennifer said, they took subcontracted jobs from her former employer. "Nobody really thought about growing a business — it sort of was like doing the work that we liked to do, and we had established somewhat of a reputation, so we were getting work that we enjoyed doing. We did a lot of work for T&P and we got to work on specifically the Nike account for T&P, which was really great creative work, so we were really busy with that." Connections to the downtown advertising scene created the context for this nascent firm, which still consisted essentially of its founders working at night, to compete for and win a contract with a major global corporation, Nike. That contract soon enabled significant expansion of the firm. "We wanted to take the challenges, to do bigger things," Brad says.

What happened was this place we were working at had Nike as a client. And Nike came to them and said, "We're opening Nike Towns all over. There's one in Chicago, there's one in Portland. We're going to open one nationally in every city in the United States. And after that, we're going to open them overseas. We need to create a campaign that announces that we're coming to town in those different cities, and that drives people to come to that store to see about working for us. What can you do for us? It needs to grab people. It needs to say that Nike feels good, it needs to say that Nike has energy to it, it needs to be exciting, see what you can do." So this company developed three different teams to develop a concept for them, and Jen and I were one of the teams. They outsourced it to us, because we had proven ourselves creatively. And we came up with the concepts, and the one that we did won. We did this idea of "a day in the life." We did like a film strip shot. We presented the concept of somebody working in Nike Town, their vision of their day. . . . That working at Nike was not just a job, it was part of their life, how they lived it, it was cool, that is, "We [Nike] want you to enjoy your job and have it be part of your life."

The association with Nike was the first step in an expanded client repertoire that shifted away from hip-hop acts toward global corporations that just wanted a hip-hop image. It enabled the small firm to hire workers and to open offices in a North Avenue loft that had in a previous life been home to the kind of garment production that Nike now subcontracts to Third World sweatshops. With their ad, they strove to sell potential workers on the appeal of post-Fordist service employment. Jobs at Nike Town offer neither security nor high wages, but they do have the supposed benefit of fun and hipness, providing employees an opportunity to draw their aura from the commodities to which they have proximity.[19] Working in a Nike Town arguably allows employees to be more "like Mike" (Michael Jordan) insofar as they now share with Jordan the structural position of shilling for Nike, although at considerably lower rates of compensation.

Talented young designers with formal training at a top graphic design program, Cowley and Arterbury have the further advantage of exposure to Chicago's hip youth culture, undoubtedly an asset in the production of a campaign with the appropriate aura of cool. In addition to this desirable convergence, outsourcing to them was also cost efficient, since they could hardly command the fees demanded by more established marketing enterprises. Following from this, we see that the overall youthfulness of the neighborhood's design sector produces a double advantage for capital, as it supplies workers who are exceptionally tapped in to new trends in culture and fashion — and who also have much smaller salary expectations than older professionals do. Indeed, the recruiting campaign that Boom Cubed produced on behalf of Nike, cajoling young workers to value fun and hipness in the workplace over salary and stability concerns, also depicts the choice many designers in the neighborhood experience having to make.

Given the revelations of hyper-exploitative labor conditions in offshore production plants (which Nike subcontracts along with its advertising needs), there is an unmistakable irony in an ad campaign designed to portray Nike as a benevolent employer. This fact was not lost on Brad. Nike's poor reputation as a global employer led to ambivalence on Brad's part, but not disavowal, as Nike's relentless commitment to ubiquitous youth-oriented advertising continues to make it a desirable client.

This was '94–'95 [when the relationship with Nike began], a couple of years before [sweatshops] came to light. There was not much criticism of Nike then, except that they were a super brand, and we enjoyed working with them. And the big thing then was they were going "officially global." And they needed a global marketing campaign to announce themselves to different cultures that they were expanding to. And we did two new campaigns for them, that they still use. That was a great challenge. We went out to Portland to present to them — that was a huge change for us. And it was pretty amazing that we were doing it on our own. This is one of the biggest companies out there, and they trust us with their work. It was great, it was exciting. We did that, and it wasn't long after we finished that project that all that stuff started blowing up about them, and they were getting this really negative press, and it was weird, we were doing all this work for them. They were using what we had done, but for a while after, there wasn't a lot of new work after the global campaign. That was an interesting thing for us to see. It was an odd thing for us, as well. It was a reference thing — OK, there's more to these companies than what they're going to tell you. I think there was a certain level of naiveté that was going on for us. I don't know how we feel about that. We did all this work for them, and [shrugs]. We have continued working with Nike — but I think there's a lot a good that they do too. Maybe that's the way I've justified it. I don't know if that's acceptable or not, but I've done it.

The young, creative employees at these boutiques are likely to be socially liberal, and, like Brad, most surely cringe at news reports depicting teenaged garment workers being disciplined with bamboo rods. However, they are committed to an image of themselves as creative workers, and Nike is valued as an employer because of its reputation for allowing a decent measure of creative freedom. When I asked the director of another local boutique if he would work for Nike, he responded without hesitation:

Yeah, yeah. Sure, there's controversy about the company. [But] . . . they allow artists to do cool stuff and pay them lots of money to do it. They're along the lines of a client that allows you to express your own artistic vision within the context of their brand. They're a brand that trusts the artist to represent them appropriately as opposed to saying, you must represent us in this very precise vision, and you've got about one millimeter of wiggle-room. There are certain clients that are a lot more flexible than others. Those clients are a lot more fun, and we do a lot of our best work for those clients.

The relationship between the corporate behemoth Nike and the tiny boutique shop in a hip urban loft illustrates the new spatial links and displacements of contemporary capitalism. The congealed value of commodities is produced at multiple sites of global production, including both sweatshops and the loft offices of marketers. The efforts of corporations to maximize their flexibility in the global marketplace have led to extensive subcontracting of activities ranging from production to marketing. The de-skilled production of Nike's material products is outsourced to offshore plants where old-style Fordist labor practices can be circumvented and the corporation can wash its hands of the brutal means by which sweatshop toil is secured. Moreover, just as this labor is devalued, the material products become displaced as the locus of corporate value. Naomi Klein suggests that corporations like Nike have "made the bold claim that producing goods was only an incidental part of their operations, and thanks to recent victories in trade liberalization and labor law reform, they were able to have their products made for them by contractors, many of them overseas. What these companies produced primarily were not things . . . but the image of their brands."[20] Image production is also something that corporations like Nike may outsource, in flexible relationships with a variety of small and large firms. These abstract relations shape the subjectivities and experiences of workers in Wicker Park, who both profit from and embody fundamental social and economic contradictions.

The Aestheticization of Information

Initial expectations of the Internet appear to have missed the mark, both in terms of its profit potential and its spatial consequences. The Internet did not rewrite the rules of capital risk and profitability the way that over-heated speculators had hoped during the heyday of investment frenzy. What the Internet has done is to join with the expanse of other media to enable the circulation of an enormous proliferation of images through space, bringing commerce and advertising more intimately into spheres of everyday existence. These images flow at previously impossible speeds through cyberspace, collapsing geographic constraints; however, they come from places, and their place origins are inscribed in the content of the mediatized global environment. The Internet and the expansion of such "venerable" media as television demand steady streams of aesthetic content, and drive new contexts of production.

The relationship between the cultural production of artists, musicians, and writers and the design-intensive enterprises of the "new economy" information technologies thus bears further scrutiny. In addition to developing as leading centers of technology over the past decade, Seattle,

Portland, Austin, and San Francisco also featured innovative youth subcultures that achieved varying levels of national recognition. The technology sector's heavy emphasis on both youth and innovation suggests that
there may be an elective affinity between "new economy" location choices
and regions that are generally amenable to subcultures and nonconformity,
a hypothesis supported by Richard Florida's "bohemian index."[21]

The celebrated "Seattle scene" produced several of the most popular
musical acts of the early 1990s, and the appreciation for such popular
culture on the part of tech workers is evident in Microsoft co-founder
Paul Allen's opening of a lavish, technologically sophisticated Rock 'n Roll
museum in the city.[22] Highly educated and disproportionately young compared to workers in other sectors, workers in high-tech industries are likely
to be culturally curious and attracted to products that are considered to be
hip and innovative. Still, if the image of tech workers increasingly came to
include excursions to alternative rock clubs in rare off-hours, there persisted
an intraregional division between the spaces of technological innovation,
which were suburban, and the neo-bohemian sites of cultural production,
which were in center-city neighborhoods.

But increasingly during the 1990s, new technology enterprises of a
specific sort began to colonize the older center city, despite the higher costs
of doing so compared to the suburbs. High-tech enclaves on the periphery
of major metropolitan areas continue to be important generative milieux
for the information economy, but for particular enterprises in new media
production, the hipness of older center-city neighborhoods is itself a key
factor of production. As the 1990s wore on, what became clear was that
the Internet was generating an incredible new demand for engaging content, both text and images. The Internet entailed not only an explosion of
information, but also the *aestheticization of information*.

The use of digital technology to generate arresting images is apparent in
a variety of media, including the motion picture industry, for which digital
imaging is crucial to the new generation of effects-driven films. New media
do not pose a technological challenge only for engineers; increasingly the
challenge is in the area of aesthetic content. Yet the Web's aesthetic content
can be, at times, indistinguishable from newly elaborated brand identities. As Klein puts it, "Every brand with a web site has its own virtual,

branded media outlet."[23] To maximize the potential of this new media, it was necessary to devise strategies to arrest consumer attention, which required both technological savvy and visual creativity. Although every major corporate interest in the United States has a Web presence, such content is not something most corporations wish to produce in-house. Internet design, a direct extension of graphic design, becomes another producer service subcontracted by corporate interests in the webs of flexible production characteristic of late capitalism. Under these conditions, the artists inhabiting the local subculture become more than just an amenity to entertain young engineers on the weekends. The artists themselves form an available labor pool with the aesthetic competencies required to meet this new demand.

As Eva Pariser notes, "Contemporary artists have embraced the Web, creating websites as a natural extension of their artistic output."[24] By the middle of the 1990s, there was an explosion of Internet use by artists who prided themselves on their ingenuity in creating websites for themselves and their friends.[25] In Wicker Park, Lumpen publisher Ed Marszewski — known there as Edmar — was among the creators in the ambitious, though currently stalled, website supersphere.com, which combined high-end graphics with left-wing politics. In 2002, he was the force behind Version >2, "the digital arts convergence" staged at Chicago's Museum of Contemporary Art.[26] The indefatigable Krystal Ashe, between bartending at Mad Bar and organizing live poetry events, taught herself html in order to help set up websites for local artists, providing freelance service at deep discounts.

The ability to harness the aesthetic potentials of digital technology remains in demand — both for motion picture effects or website design — and gives young artists a new means of converting their creative competencies into remunerative employment. As Michael Weinberg, Buzzbait's founder, told me, "My friends used to say there's no money in art. I say bullshit. I'm making Web pages, full-motion video, and, it's true, Digital Kitchen [a digital-effects firm] can't scrape up enough visual-effects people here [in Chicago]."

Work relationships in the design sector blur distinctions between the bohemian lifestyle favored by artists and the capitalist workplace. Weinberg in-

dicates that individuals who may have languished on the bohemian margins
in the past now are inserted into the flexible relations of new media production. "Lots of people who fell between the cracks in another generation and
were more marginalized are [now] highly employable and catered to, and
the businesses tend to be flexible with their lifestyles and life cycles. Flexible
workdays from home, countercultural environment in the office. Gaming
companies, animation companies, broadband contact creation houses."
Employees of media-design firms are encouraged to consider themselves as
artists regardless of whether their efforts are for display in outré galleries or
made on behalf of corporate clients. Artistic experiments with the new media
effectively bleed into more profitable enterprises, developing a digital sector
that has evolved in response to the contemporary demands of capital. Thus,
a *New York Times* article noted that the "new media" in New York began with
"artsy, underfunded Web sites . . . [but] the new media seem to be maturing
into on-line extensions of the old-line Manhattan media — whether in advertising, marketing, publishing or broadcasting."[27] Firms in Wicker Park
show a similar evolution, although their maturation no longer ends in a
successful IPO and early retirement. Streams Online Media Development
began with hip vanity projects like the popular (but not profitable) interactive Web game "Piercing Mildred,"[28] and, as we saw, Boom Cubed began
with its founders' making fliers for "underground" musical performances.
Both would shortly use their hip aesthetic sensibilities in the service of large
corporate clients.

In their study of the "geographies" that organize global capitalism,
Scott Lash and John Urry observe, "The spatial dispersion of the vertically (and functionally) integrated firm, most significantly on a global
scale, occurred at the same time as the growth of vertically disintegrated
economic districts in the major agglomerations."[29] These districts encourage freelance employment and ad hoc collaborations among individuals
and small firms. The do-it-yourself ethos of bohemia supports forgoing
employment security in exchange for a feeling of autonomy, including the
freedom to pursue other arts interests. Buzzbait's Weinberg indicates that
"some of those people can also be independent vendors in commercial
art services, not necessarily looking for an organization but starting ones
of their own or working as freelancers." This allows local firms to expand

and contract their workforce with relative ease as conditions dictate, a key aspect of post-Fordist employment strategies.

Despite the technological component involved in filling the image demands for new media, these firms are perhaps best understood as flexible appendages to Chicago's already substantial advertising sector. In 1995, results from a study titled "Geographic and Political Distribution of Arts Related Jobs in Illinois," commissioned by the Illinois Arts Alliance Foundation, demonstrated that the advertising industry was the leading source of arts-related job creation in the core city area.[30] Many of the workers in the boutique firms located in Wicker Park have formal training in graphic design and previous experience working with large advertising agencies either as salaried employees or as freelancers. Ties to the established advertising sector downtown are crucial to the development of the neighborhood media sector, as prominent advertising firms often mediate linkages between the boutique shops and corporate employers seeking hip new images, as was the case with Boom Cubed's work on behalf of Nike.

Local Talent

Dave Skwarczek, founder of Streams Multimedia, said that Wicker Park was an advantageous site for labor recruitment as the firm expanded throughout the decade. Although the firm was always focused on Internet services, technical savvy was viewed as less of a prerequisite than creative skills. The following excerpt from an interview I conducted with Skwarczek illustrates these points:

RDL: Was it hard to find talent?

DS: It was not hard to find creative talent. It was hard because we ended up having to teach them how to do the Web stuff. Now, there are people that can do those things, so you can find people with some understanding of the Internet.

RDL: Did you find in those early years that you were feeding off of the neighborhood, and if so, how would that work?

DS: I remember this one artist who I was working with at the time. We've got a lot of her painting around here. She was a really fantastic and really prolific painter. . . . I was really attracted to her talent. First of all, she was a really cool person; and secondly, she was really, really ambitious. I mean, I know a lot of painters; and maybe they would work on one painting at a time. She would have five. . . . She was just oozing with creativity and really has her own style. Even today she still has her own strong style, and it's kind of evolved from what she had at the time. She was just one example [of a local artist we hired]. Some of the folks that we hired throughout the years lived around here because there were always a lot of really cool people that were really ambitious and talented . . . and whenever we had opportunities, those were the folks that we turned to. I think that the neighborhood kind of meets expectations by not meeting expectations. People expect the creatives to have long hair and paint on their clothes and talk like beatniks and recite poetry. They expect that from people that provide creative services. That's what this whole area is about: People that come here to express themselves, and are expressing themselves, and meet other people as different as they are. . . . There are all kinds of things you get here that you wouldn't necessarily get in a more homogenized atmosphere.

The ethos of nonconformity engendered in the neighborhood, "meeting expectations by not meeting expectations," in fact produces individuals with the dispositions and competences necessary for neighborhood media enterprises, avatars of hip consumerism.

Managers at these firms are likely to be genuinely sympathetic to the desires of employees to maintain diverse creative interests. Employers of one young designer, who currently plays in a punk rock band that has recorded with an independent label and toured internationally, have allowed him some job flexibility so as to accommodate him in his musical career. This flexibility is one advantage of working at a small neighborhood firm, though the advantage comes at the presumptive cost of the higher wages and increased upward mobility that might be available in a more corporate environment. For this designer, the trade-off is worthwhile.

That's the cool thing about Boom. Whenever I had to go on tour to Europe for three weeks to a month, they'd let me go, as long as I told them in advance. They're really forgiving about that. I think that I'm probably not making as much money as I want to, because I have the freedom to do what I have to do with my music, and it's sort of a compromise I have to make. I could have gone out to look for a job somewhere else, but one of the reasons [that] I didn't is because I know it would probably be a tighter and more corporate culture, and I wouldn't have as much freedom to do things I want to do.

Although this designer is commonly assigned projects outsourced from large advertising firms to promote corporate products, he is nonetheless able to experience himself as retaining a measure of professional autonomy and as being insulated from the stultifying culture of bureaucratized corporate organization. And although the freedom of his present employment arrangements is purchased with the presumptive sacrifice of higher wages and future upward mobility, the availability of such remunerative work in a more "corporate" environment appears to be a matter of faith rather than experience. The break with Fordist relations of employment, with an ethic of reciprocal loyalty between employee and firm replaced by more precarious arrangements, is culturally absorbed into what Ulrich Beck describes as "the paradoxically collective wish to live a 'life of one's own.'"[31] Thus, workers such as this one can feel as if they have willfully rejected work arrangements that are in fact increasingly scarce in the age of downsizing and flexible accumulation. The feeling of autonomy belies the fact that their labor still ultimately operates in the service of large-scale capital accumulation. Beck adds, "They work on their own life as if it were a work of art, yet have to obey the dictates of competition and global corporate power."[32]

In many cases the "boutique" founders themselves do not make wages significantly higher than their employees, and they may even make less while doing similar types of work. Their encouragement of employees' diverse interests and their offer of flexibility to accommodate them is not a matter of simple employer benevolence. This flexibility is part of the bargain that helps keep designers employed at firms that do not generate revenue sufficient to pay large salaries. Moreover, employers are conscious that active participation in the local arts scene actually increases their employees' value insofar as it promotes the further development of creative competencies. The following excerpt, from an interview with the director of a local firm, shows that he both identifies with his employees and recognizes the benefits that come with encouraging outside creative interests:

> Well, like I've said, I've always wanted to create a place where I'd want to work. The place where I'd want to work would support my creative endeavors, and the kinds of creative things that I did on the side, and would recognize the fact that if I was continually building my skills with my own stuff, it would also in turn benefit the company. That's the one thing that I really hope everybody does while they're here — learn stuff, get better, and realize their own dreams as creative people. Everybody here is really creative. Everyone has their own vehicles for expressing themselves. I can only hope that those guys are half as happy doing that kind of stuff as I've been.

In the neighborhood design sector, employees and freelancers are often encouraged to think of their design work as an extension of their artistic personas. Andrei provides an example of this disposition, citing diverse sources of inspiration in both his music and his design work:

> I never — well, I used to think of myself as a musician or designer, but now I just like to think of myself as an artist because, I don't know, I never did like the idea of pigeon-holing myself into one category. Lately, for example with the design, when I go on the Web or when I see design work, it seems to draw influences from other things like glass blowing, 3-D video game rendering, and video art. If I were to consider myself as a designer, I would probably restrict myself to the whole 2-D aesthetic you learn in design school, positive/negative space relationships. . . . I think design is really locked into one school of thought. I don't want to do that.

I've always more been into things that are outside of what I know, like glass blowing, video art, and especially working here I've been exposed to that. We have other people here who come from those backgrounds, and seeing that stuff is really, really cool, and I definitely think it has a positive effect on my design *and* my music.

To the extent that employees in Internet and graphic design experience their work as of a piece with their overall artistic personae, they experience tremendous pride in and personal commitment to the finished product. These dispositions support intense creative efforts, but they also create real issues of employee morale when the expectations of clients diverge from the vision of the designers. Despite a work culture that makes numerous concessions to the idea of personal creativity and autonomy, work in the design sector is still wage labor, and creative freedom has its limits. As one Boom Cubed designer puts it:

There are times where we'll run into clients who want something specifically, and we have to give them what they want. I'll be honest: It's kind of frustrating because you tell them it's not going to look good this way or that way. They tell you, "We're paying for it, we want it that way." Then they get what they want, and they come back and say, "We don't think it looks so hot." It's like, well, if [they] would have listened to me, we could have avoided wasting all this time and energy, and we would have gotten something that looks good. There are times where there isn't freedom.

Several firms in the neighborhood began with clients from the neighborhood, such as bars or bands, who paid little but whose needs were consistent with the creative self-image of designers invested in local culture. However, the growing reputation of the neighborhood design sector helped these firms win corporate contracts that were more lucrative but more mainstream. Taking on large corporate clients makes it more difficult to ignore the contradictions between creative freedom, personal ideology, and business necessity. At Boom Cubed, the increase in internal emphasis on marketing alienated Andrei:

I knew that design was a form of communication, and I know that marketing is a form of communication, but I never thought of myself as some sort of subsidiary to marketing campaigns. . . . It was like a slap in the face. It was this person coming in and telling me that he's going to articulate these concepts and ideas, and I would have to execute them. I didn't take it lightly, and I let him know that. There was a lot of tension at first. It was really hard to have a marketing guy come up with a color system or branding, the way things would fall place and how design would be applied to that. Tag wise, coming up with designs that would revolve around that was a really hard pill to swallow at first. Also, the fact that they had these clients that they wanted to shoot for, and I wanted to be a part of a design culture that was more in line with subculture, stuff like skateboards, snowboards, BMX bikes, record labels, just stuff that was a lot more fun designwise. They wanted to do stuff like McDonalds and Coca-Cola. It was really hard. It was hard to soak in and gel with.

The frustrations of client compromise are issues that firm directors must deal with. As the previous passage illustrates, creative individuals in the neighborhood are often not tuned in to the bottom-line realities of business, and can become petulant when these issues intrude on their creative visions. A manager's sympathy usually lies with the designer and not the client, especially since firm founders are almost always directly involved in the creative process themselves. However, sympathy does not pay the bills. Since they have much more at stake in overall firm performance, directors reluctantly accede to the awareness that clients' desires must be satisfied to maintain business viability. Local firms respond to this problem with limited provisions to pursue personal interests on company time a strategic means to overcome employee resentment and keep creative commitment high, a tactic that at least gives employers a chance at the moral high ground. Buzzbait's Michael Weinberg explained in our interview:

MW: There's a constant tension with us [about] our designers doing what they want to do — following their creative vision on something — versus what the client wants, and that is a constant tension in this kind of field. It really is [a problem], trying to rein those guys in, and there's a lot of prima donnas, there's a lot of competition between them or among them of focus of vision, but also a lot of resistance to meet commercial requirements to sacrifice at the cost of vision. It's a constant reality.

RDL: How does that get resolved?

MW: Different ways and different places. With a bitch slap from me. I'm coming down more on that. There are lots and lots of free-form creative outlets that we make for people to do here, and they really don't follow those up. Plus you have your time to read this and that, do an alternative site for us, [put up] a showcase for stuff for you, [make] T-shirts, do print work here, pro-bono stuff for bands or for whatever we want, and they don't do it, and then they want to bitch about that. *They have to do what the client is paying us to do.*

Jennifer Arterbury likewise recognizes the inevitable frustration of creative compromise, and suggests that the diverse outside interests abetted by the neighborhood scene provide a useful tonic:

I think it's really important to go out and be inspired by other things. I don't think you can sit here and design all day without having outside interests. Design, although it is creative, it is considered an art form depending on the project, but it's not autonomous. You're always being influenced by your client, and there's a lot of compromise. That's just the way design is. In turn, a lot of people who are designers who have to change what they've done, or [who] really like the work and the client may not be happy with it or wants it changed or doesn't like it at all, and they have to change it, it can be really frustrating because they feel really good about that work, and it's somewhat personal. They invest time and energy, and then you have to change it for a reason that may be a good reason or may be a bad reason, but in turn that creativity is taken away so you really need to have something you like to do. That goes with any job, but for designers especially it's good to be involved in the music scene or going to galleries or whatever.

Boom Cubed actively encourages employees to avail themselves of the city's cultural offerings, maintaining a company membership at the Art Institute and the Museum of Contemporary Art available for use by all employees. This nurtures the sense of continuity between design and the fine arts. The identification of local designers with their work as an extension of other creative interests is useful, motivating commitment and easing tensions. But the fact that these design efforts are ultimately made on behalf of corporate interests entails compromises that cannot entirely be obscured and that many designers are inclined to resent.

Further, local dynamics of gentrification and concomitant change concern employees, who view neighborhood diversity as a key to their work environment. Boom Cubed is particularly geared toward multicultural aesthetic production. The currently multiracial, multiethnic (but uniformly young) staff of marketers, programmers, and graphic designers are interspersed throughout Boom's open loft offices on North Avenue. As Andrei describes it:

> Internal culture is pretty mixed, you got people who are into hip-hop, you got people who are into R&B, dance music, people who are designers, coaters, people who are marketers, people who do events, people who do administrative stuff. It's really diverse; even race-wise, it's kind of cool. This is probably the first company I've been at where you have a mixture of Latino, Asian, African American, Caucasian. I think that's cool. All my other jobs it's been predominantly Caucasian. I was the only Asian there.

According to Arterbury, cross-cultural interaction is a feature of the neighborhood spatial practices:

> For instance we had an event last week at Pajé, which is down the street on Milwaukee. That restaurant is African American owned

[and] is all southern food, and the event was interesting because I expected mostly just an African American crowd, and initially it was, but it was more diverse, and I think [that is] somewhat because of the neighborhood. A lot of different types of people showed up, whereas if it was somewhere else, I don't necessarily know if that would be the case. I found that, this is just observing, just for instance, the open mike night that goes on at [neighborhood bar] Pontiac is very diverse. It's pretty much the most diverse open mike I've ever been to. As far as ethnicity goes, they are much more cross-cultural. I think that people just working here, in this area, Shannon is [white] female, she opened this restaurant, her husband Malcolm is African American, the people who work there, they are all completely different people. There's not one type of people there. When you go there to eat, there's not just a white crowd there, not just a black crowd, it's all different types of people who enjoy it. You'll find older people there. I like that. That's just what I see in this area right now.

At the same time, Arterbury worries that this neighborhood balance is fragile, and potentially vulnerable to the homogenizing tendencies of gentrification. She is distressed by the recent opening of Starbucks on the central six-cornered intersections of North, Damen, and Milwaukee Avenues. "It makes things boring. Once you have the same stores in every neighborhood, who cares? The same neighborhood I can go to in the suburbs." Her fears echo those of critics who see the city being transformed into a theme park by soulless capitalist interests who undermine place identity with generic consumption offerings, of which Starbucks is only the most familiar example. As rents increase, it becomes harder to sustain nonprofit organizations like the Guild Complex or venues such as independent galleries with cultural attractions that are more esoteric.

Focusing on the neighborhood culture as an amenity can give only a partial account, however, if one understands it as a story mainly about consumption. The neo-bohemian place identity itself serves as a factor in production, directly inscribed in design products. Moreover, the neighborhood does not just concentrate talent but also nurtures it, while enabling the social networks and work habits that characterize flexible relations of production. The artistic subcultures of the neighborhood magnetize individuals who may possess appropriate aesthetic skills. Beyond the simple concentration of labor, the local culture contributes to the ongoing maintenance of hip cultural capital and creative competence.

Conclusion

CHAPTER 10

The Bohemian Ethic and the Spirit of Flexibility

I had reservations about making art a business.
But I got over it.
  —Mary Boone, SoHo gallery owner, 1982

BOHEMIA HAS RECEIVED CONSIDERABLE ATTENTION in recent years, in both scholarly works and entertainments directed at high and middlebrow consumers, with major revivals of *La Bohème*,[1] the success of the Broadway musical *Rent*, an Academy Award nomination for Baz Luhrmann's film musical *Moulin Rouge*, and the release of smaller films including *Joe Gould's Secret* and *Beat*. This should not surprise us; as Wendy Griswold argues, revivals say as much about the contemporary period as they do the past.[2] Examining Wicker Park illuminates why fantasies of *la vie bohème* resonate with contemporary cosmopolitan audiences, if not with some mythical mass culture. This contemporary district in Chicago updates many of the thematic principles of classic bohemia; indeed, participants often make explicit recourse to past bohemian examples in their contemporary designs for living. Nor is the Wicker Park case eccentric; similar districts have become more frequent, more visible, and more important to understanding the culture and economy of contemporary cities around the United States.

Beyond even this, though, are changes in the contemporary economy broadly, with an enhanced emphasis on both individual creativity and the individualization of risk. Compared to a previous era, flexible capitalism demands greater adaptability from its workers, and even educated professionals must learn to live with contingency and vulnerability. In other words, the reality of their work lives pushes them closer to the lived experience of urban bohemia in key respects. Thus, while David Brooks argues that BoBos maintain bourgeois orientations to work, taking only their consumption cues from bohemians, in fact the contemporary resonance of bohemia penetrates more deeply than just a fashion statement. New-economy professionals are not bohemians, but they are not "organization men" (or women) either, and in this period of neoliberal capitalism, it may be the bohemian ethic, not the Protestant ethic, that is best adapted to new realities.

Structural Nostalgia

Despite, or perhaps because of, its prominent place in the popular vernacular, bohemia remains a tricky concept to pin down, used in a wide range of loosely affiliated fashions. I have attempted to harness it by emphasizing bohemia's material, place-based attributes as well as its cultural connotations, and by anchoring my analysis in a specific case example. Still, while elaborating this project over a number of years, I have encountered periodic resistance to my application of even a modified concept of bohemia to Wicker Park. Principally, skeptics make recourse to the loaded concept of authenticity, essentially asserting that the facile efforts of contemporary Wicker Park artists do not warrant so distinguished a pedigree. A smaller cadre, usually drawn from a younger audience, argues in the other direction, claiming that today's hip urbanites are something brand new, having transcended the tired ethos of bohemian alienation. Both objections are profoundly value-laden, and arguing them is useful only up to a point. I have been at pains to demonstrate that there is sufficient continuity between Wicker Park and the traditions of the urban bohemia to make the term a useful conceptual (not normative) point of departure, even as I have affixed the prefix "neo" to highlight the important ways in which contemporary bohemias are historically distinct from their predecessors.

Still, even if prepared to concede the efficacy of "bohemia" as a characterization of some recent period of Wicker Park's past, many will argue that this is all over now. In fact, by the time I began making the neighborhood rounds in the fall of 1993, it was already a commonsense assertion locally that the neighborhood's best days as a bohemian enclave were already over. In the time that I lived and/or worked in the neighborhood, a span of some eight years, numerous events were identified as signaling the new bohemia's ultimate demise: enhanced corporate sponsorship of "Around the Coyote," the closing of Urbus Orbis (or the Busy Bee, or the Hothouse, or Mad Bar, or whatever), the opening of Starbucks, the arrival of MTV. Obviously much has changed, and I myself experience almost unbearable bouts of nostalgia on return to the neighborhood streets. For all the new development, though, Wicker Park retains a good deal of lively and funky ambiance and much of the vaunted gritty feel of days past.

The sense of being always already over is an apparently structural feature of both classic bohemias and their contemporary heirs. As Ross Wetzsteon writes in his history of Greenwich Village:

> For all the rhapsodizing about "happy days and happy nights," from its very birth, bohemia seemed to exist in the past. "Bohemia is dying," even its most ardent residents lamented; "the great days of bohemia are over." This sense of lost grandeur has existed in every generation — just as Floyd Dell said in the teens that "the Village isn't what it used to be," Murger's followers were saying in the 1850's that "Paris isn't what it used to be." "Whatever else bohemia may be," a Village magazine editorialized in 1917, "it is almost always yesterday."[3]

In Jonathan Larson's rock opera *Rent*, the *La Bohème* of the early 1990s East Village, the first-act climax features the lines: "Bohemia! Bohemia is a fantasy in your head. This is Calcutta, bohemia is dead."[4] Likewise, anyone who has lived in Wicker Park for more than a few months feels entitled, and perhaps obligated, to insist that the neighborhood was much better in the old days.

What is the source of this structural nostalgia? It is not reducible to the life course trajectories of participants, confusing lost youth with lost bohemia. Since it recurs incessantly, bohemia dying a thousand deaths, it cannot be only the product of some objective change, like the closing of Urbus Orbis or the opening of Starbucks. Rather, bohemia is always already over because it always already falls short of its adherents' fantasies of social autonomy, expressed in the vaunted ideology of *art pour l'art*. From their very beginnings, bohemias and neo-bohemias are subject to external and internal pressures, pressures that differ in each historical period; thus the perpetual nostalgia for an imagined moment of genuine independence (and efficacious opposition). And, as Michael Herzfeld argues, "the static image of an unspoiled and irrecoverable past often plays an important part in present actions."[5] The continuously revived image of an elusive Edenic moment — the "good old days" of bohemia — in this manner becomes itself part of the ongoing representations, and conflicts, within contemporary Wicker Park.

Bohemias old and new are nested communities, embedded in initially poor or working-class neighborhoods where the bohemian participants are a minority of the overall population. As subcultures they are defined in relation to an increasingly diffuse parent culture.[6] Problems of coherence, membership, and authenticity are present virtually at the creation of these entities. These problems have only grown more intense in contemporary contexts; artistic subcultures animate the fantasy lives of young people now more than ever, and even urban professionals with college educations agitate for participation. This tendency increases the marketing advantages of new bohemian spaces in the city, while complicating boundary maintenance for committed scene makers.

But if more complicated now, the activities that constitute bohemia have nonetheless never stood apart from the broader forces of economy and society. The Parisian prototype cannot be understood without taking into account the structural upheavals of nineteenth-century modernity — the explosive growth of the metropolis, the ascendance of the bourgeoisie, the submission of artistic output to new market forces. The beats of the mid-twentieth century (to take another example) similarly elaborated their artistic and performative innovations in the context of postwar Fordism, with its unprecedented expansion of the middle class, the growth of the suburbs, and the correspondence of mass standardized production with a social ethic of mass consumption. Neo-bohemia, for all that it derives from the examples of bohemias past, is distinguished from them by its own structural contexts, today associated with globalization, neoliberalism, and the postindustrial metropolis. Bohemia has traditionally been considered marginal and subversive within the capitalist economy that nonetheless called it into being, a position that was probably always overstated. Today's bohemians have not necessarily abandoned that stance (despite Richard Florida's claims to the contrary), but as a practical matter Wicker Park's neo-bohemia enhances profit-generating strategies in a variety of new, and occasionally surprising, ways.

Wicker Park's neo-bohemia thus confounds traditional conceptions of urban subculture, which offer the alluring image of a counter-hegemonic resistance to capitalist domination.[7] While proponents of this view typically concede that such subcultural innovations may be co-opted by capitalist interests in arenas like fashion and media (in the process robbing them of their subversive intent), this is presumed to happen after some more pristine moment.[8] But this division does not capture the actual fluidity of the boundaries between the articulation of cultural innovation and strategies of accumulation. Rather than looking at artists as a resistant subculture, I became compelled to think of artists as useful labor, and to ask how their efforts are harnessed on behalf of interests that they often sincerely profess to despise.

The New Bohemia and Flexible Labor

As in past bohemias, contemporary participants in the Wicker Park scene insist upon their opposition to an imagined mainstream. When elaborating upon this objection, local artists specifically repudiate participation in a corporate workforce committed to conformity and the base pursuit of material security. Yet this imago of the mainstream is anachronistic, as the old promises of career and social security under the terms of the Fordist corporation and the welfare state have increasingly evaporated. As Vicki Smith notes, "Uncertainty and unpredictability, and to varying degrees personal risk, have diffused into a broad range of postindustrial workplaces, service and production alike. Tenuousness and uncertainty have become 'normal' facts of work and employment across the occupational spectrum in the United States."[9] In addition to requiring that workers acclimate themselves to greater flexibility, with volatile compensation and irregular work schedules, the flexible workplace makes increasing demands on the individual's creative capacity, even in mundane service-sector jobs. Writes Ulrich Beck:

> Never before has individual creativity been as important as it is today. . . . But never before have working people, irrespective of their talents and educational achievements, been as dependent and vulnerable as they are today, working in individualized situations without countervailing collective powers, and within flexible networks whose meaning and rules are impossible for most of them to fathom.[10]

Today, workers must be competent to the task demands of flexible production — able to demonstrate "individual creativity" to an unprecedented degree — and they must also be able to acclimate themselves to enormous amounts of uncertainty and risk.

The "ethical" dispositions nurtured in the bohemian milieu (but not confined there) may indeed have been incompatible with the highly routinized labor that prevailed in the Fordist city, the world of the assembly line worker, or the other directed "organization man." But *la vie bohème* has long been characterized by the insecurity that now infects broad swaths of the postindustrial economy, and we have seen that the bohemian ethic elevates tolerance of uncertainty to a virtue: "As an artist you know that you many not be secure for the rest of your life." And the disposition to wear risk as a badge of bohemian honor is carried by neighborhood artists even into non-art employment.

Consistent with tradition, what lends this gesture its heroic aura is the conviction that one is consciously rejecting a more secure, conventional, and stultifying life. Recall Andrei, an Internet designer/punk musician, indicating, "I could have gone out to look for a job somewhere else, but one of the reasons [that] I didn't is because I know it would probably be a tighter and more corporate culture, and I wouldn't have as much freedom to do things I want to do." But the corporate culture to which he refers is derived less from current reality than from the inherited image of the "organization man" that still animates bohemian ideology despite its increasing inadequacy. Rejecting such labor in the 2000s is a gesture very different from what it was in the 1950s, since to a large extent it doesn't exist anymore. In this broader context, the bohemian disposition that makes "living on the edge" a supreme virtue is in fact quite adaptive to labor realities.

Against the antagonism presumed by Daniel Bell in *The Cultural Contradictions of Capitalism*, what I am suggesting is an "elective affinity" between the dispositions enshrined in bohemian ideology and the requirements of select new capitalist enterprises. By maintaining that the "techno-economic realm" of society remains essentially bureaucratic and hierarchical,[11] Bell failed to adequately anticipate the consequences of the postindustrial society whose arrival he was forecasting in the early 1970s.[12] He therefore did not see that rejecting "sober bourgeois capitalism" or routinized mass production does not necessarily mean rejecting capitalist labor altogether. The production of art has always entailed frenzied jags of activity. Indeed, both Balzac and Murger hastened their deaths through the copious ingestion of caffeine, fueling all-night writing sessions. Is this so different from software designers, similarly stoked by what is today the signature beverage of yuppie overachievers, crunching to make deadline? Is Kerouac's roll of butcher paper, used to compose *On the Road*, so different from the continuous scroll of the word-processing program that I am using now? Past styles of capitalism may not have been able to harness the excesses of creative energy in bohemia beyond the limited cultural marketplace, but contemporary capitalism, accentuating design, fashion, and flexibility, exploits these energies in myriad ways.

The education and diverse cultural competence of young artists further increases their usefulness in a changing urban economy. The traditional do-it-yourself ethos of bohemia fits in well with the entrepreneurial imperatives of neoliberal capitalism. Not surprisingly, enterprises eager to exploit these qualities gravitate to the neighborhood. It is this confluence that sits behind Richard Florida's imagined resolution of bohemian alienation. Though bohemians once patrolled the margins of society, Florida insists that the creative class sits today at the center of a new mainstream, happily embracing the risk society, at least until something actually goes wrong. Wicker Park's technology sector presents a less rosy picture, one in which opportunity is more limited, compromised vision more common, and fantasies of spectacular wealth increasingly no more likely to materialize than are dreams of becoming the next Cobain or Basquiat. Moreover, whether or not these creative workers are willing to acknowledge it, they are haunted by the ghosts of bohemia past as they deploy their efforts in a new urban milieu. The traditions of dead generations are what make it possible to understand oneself as resisting the stultification and injustice of corporate capitalism while working twelve-hour days making recruitment ads for Nike.

Winners and Losers

Throughout the 1990s and beyond, Wicker Park concentrated a large number of talented, ambitious individuals, and no doubt an even larger number of dabblers and hangers-on. To varying degrees, these participants could imagine that they would eventually join the ranks of the cultural elect, winning fame and fortune for their creative output. From our current vantage point, we can easily slot them into the appropriate categories: those who succeeded in the arts were talented and ambitious, and those that didn't were dabblers and hangers-on. After all, reading backward creates the illusion of career inevitability, but reading forward is a far trickier matter. The truth is that even with its dramatic expansion, the cultural marketplace provides a limited amount of opportunity, not nearly enough to accommodate even the very serious and very talented. Moreover, hardly anyone believes that this market is an efficient or transparent mechanism for sifting out the worthiest individuals and products, any more than there is local consensus on who is worthy and who is not.

It is only to be expected that most associate bohemian scenes with those whose creative profile is the highest, so that Jack Kerouac exemplifies the beats and Kurt Cobain exemplifies grunge. But these individuals are necessarily a tiny minority of participants, and overcredited in retrospect with making the scene. To focus only on the successful artists obscures the fact that even the poseurs and dilettantes play an important role, showing up at performances, gallery openings, and loft parties, and thus ensuring that there is a scene at all. In fact, self-absorbed local talents often free-ride on the efforts of the other locals that do the work of organizing shows and assembling congregations. As with Mabel Dodge or Sylvia Beach in bohemias past, the hosts of the salon remain a core, if underappreciated, component of the collective enterprise, without which Marcel Duchamp or James Joyce might well have had nowhere to go.

The work that goes into maintaining the local scene benefits artists and also a variety of other economic interests. Culture-industry gatekeepers follow the crowd of tuned-in consumers in the search for new talent. Though unpaid for their advance work, Wicker Park locals are useful as avatars of cool. Moreover, as has been stressed over and over, the cultural marketplace is no longer the only, or even the primary, way that the new bohemia intersects with the urban economy. The paucity of direct economic returns does not make the arts unimportant to the local economy; rather, this importance is complex and mediated.

As I've shown, participants typically bring with them unusually high levels of education, or human capital; as they interact within the neighborhood, they further increase their stocks of another sort of capital. Knowing how to dress, how to evince the appropriate demeanor, how to talk knowledgably about various non-mainstream cultural offerings — these are skills honed in the local milieu, and are attributes that can be conceived of as subcultural capital, though only in specific contexts. What makes competence into capital is the opportunity to leverage it for social rewards. Under the right circumstances, subcultural capital can confer privileges of status and money. Artists have many more ways than in past periods to sell their dramatic personae and hip tastes in the contemporary city, for example, in bars and nightclubs, or the media design sector. Still, the greatest profits frequently accrue to players other than the artists themselves — property entrepreneurs, venue operators, or large corporate interests like Nike and MTV.

Antonio Negri argues that in the contemporary period of capitalist development, "work processes have shifted from the factory to society, thereby setting in motion a truly complex machine."[13] But while creative labor may resist the routinization of the factory floor, it still occurs in actual places. The practical interactions in these Wicker Park venues shed light on how this "complex machine" operates in concrete locales. Art and the artist's lifestyle are organizing principles, giving coherence to the local scene, but the new bohemia generates value in a wide variety of ways. Regardless of their talent, participants are seldom able to sustain themselves through pursuits like acting, performing music, writing poetry, or painting; nor do these pursuits in themselves generate much surplus value for others to extract. But even "leisure" activities in spaces like Urbus Orbis or Phyllis' Musical Inn turn out to be much more; Wicker Park concentrates what we might call productive leisure.

An analogy can be drawn here between the arts in Wicker Park and the type of plot device that Alfred Hitchcock termed a "MacGuffin."[14] Essentially, a MacGuffin forms a pretext that sets into motion the real action of the plot. While of crucial importance to the characters, its importance to the story lies in its function as a catalyst rather than in its intrinsic value. The arts serve as a MacGuffin for postindustrial economic activities in Wicker Park, a pretext for particular patterns of congregation that then prove useful for other kinds of enterprise. The paltry returns for most artistic efforts are precisely what enable the incorporation of local artists into these other labor contexts. Indeed, if fields like theater, poetry, or the visual arts had a larger market and better-compensated participants, it would reduce the need and opportunity for artists to capitalize on their subcultural competence in the more commercially viable enterprises that have agglomerated in and around the neighborhood.

Place and Cultural Production

Even with the staggering rise of suburbs and edge cities as sites of residence and economic activity, cultural production still privileges the old center city as a generative milieu and site of fantasy brokering consumer desire, particularly for consumers in the coveted youth demographic extending to age thirty-five. In the 1990s, major trends of youth fashion — including grunge, hip-hop, and heroin chic — were clearly sold as emerging organically from street culture (and each came with its own musical soundtracks and celebrity icons). The suburbs, with their strip malls and "virgin sidewalks,"[15] generate no such alluring associations. A neighborhood like Wicker Park, put on the cultural map in the 1990s by rock 'n' roll culture and its associated street aesthetic, can be a "real world" model for constructing the image of "a hip downtown scene" that becomes pseudo-universalized as it enters the global swirl of commodified signifiers.

Making the argument for neo-bohemia as a key feature in new urban economies requires sensitivity to the embedded nature of all local space in larger processes. But these global processes are only the accumulated effects of diverse localities and networks of localities. By zeroing in on the neighborhood, using the most intimate and site-specific methodologies, this study restores materiality to an economy in which the older spatial anchors, particularly the factory, recede in explanatory power, at least for the post-industrial West. The study also tells us much about the process of identity formation among the "cosmopolitan" social subjects of the global city, with local identification turning out to play a larger role than is often suspected.

During the 1970s, the decade that brought us *Taxi Driver*, fiscal crises were undermining even basic service provision. Generally speaking, it was the suburbs that were believed to have the amenities most appealing to the middle and upper classes, from galleria-style malls to the all-important good schools. But if *Taxi Driver* reflected a pervasive sense that the city had become a dangerous and disorderly warehouse of vice, by the 1990s the suburbs were under increasing attack for their own inadequacies.[16] While artists have long dismissed the suburbs as sites of culture death and conformity, this view has now extended along with the bohemian ethic to affect ever-larger swaths of the populace. And the new amenity that the suburbs are said to be so sorely lacking is precisely the sort of lively, culturally diverse sidewalk life that Jacobs gleaned in Greenwich Village, America's proto-bohemia.

For its part, Wicker Park may no longer appeal to those most keen to find the cutting edge, and they will have to look elsewhere. But pressures and contradictions within the structural field foreordained the scene's instability; neo-bohemia is not a reified natural area but rather a mode of contingent and embedded spatial practices. Like Greenwich Village, the legacy of its bohemian moment still contributes to ongoing representations of Wicker Park, patterns of enterprise and the realization of local value, even if starving artists are priced out. Undoubtedly new sites, still unrecognized by the media, sit poised to assume the mantle of "cutting edge's new capital," and undoubtedly they will also relinquish the title quickly thereafter. But the theory of neo-bohemia is about much more than the life of a single neighborhood; it invites us to rethink in broad ways the interrelations of lived space, subjectivity, and instrumental labor in this contemporary period of globalized capitalism and flexible accumulation.

NOTES

Chapter 1

1. The name derives from the title of the Douglas Coupland novel *Generation X: Tales for an Accelerated Culture* (New York: St. Martin's Press, 1992). "Slackers," another once-popular shorthand for this allegedly work-averse cohort, comes from Richard Linklater's film *Slacker* (1991).
2. Noshua Watson, "Generation Wrecked," *Fortune*, Oct. 14, 2002.
3. The Rush/Division nexus was a focal point for David Mamet's stage meditation on young urbanites in lust, *Sexual Perversity in Chicago* (1974), adapted to film as *About Last Night* (1986).
4. Actually, it would be a mistake to overstate this as a source of distinction, since many young participants in the Wicker Park nightlife were plenty interested in maximizing their sexual opportunities. Mimi Schippers argues that the "alternative hard rock" culture of 1990s Chicago was exceptionally progressive in its sexual politics, challenging the patriarchal conventions of mainstream urban singles' scenes. Schippers indicates that these hard rockers subvert gender expectations via postfeminist, postmodern tactics of "gender maneuvering." The central maneuver Schippers describes seems to be for young women rock fans to dress in a sexually provocative manner and then loudly reject men who approach them — except for men who are musicians. I don't dispute that this sort of behavior occurred all the time in such Chicago rock venues as Lounge Ax, Cabaret Metro, and Empty Bottle, but I question whether it is either as historically original or as subversive as Schippers argues. See Schippers, *Rockin' Out of the Box: Gender Maneuvering in Alternative Hard Rock* (New Brunswick, NJ: Rutgers University Press, 2002).
5. We will learn more about Algren and his fiction in later chapters. Writing in the tradition of the Chicago realists (though as we will see with important variations), Algren depicted Wicker Park as a hardscrabble neighborhood populated by junkies, prostitutes, and hustlers. Upscale condominiums now occupy the lot.
6. Eric Boehlert, "Chicago: Cutting Edge's New Capital," *Billboard*, August 1993.
7. See Schippers, *Rockin' Out of the Box*, pp. 42–43.
8. Jeff Huebner, "A City Without Art" *Chicago Reader*, Nov. 2, 2001.
9. For more on Razorfish, see Michael Indergaard, *Silicon Alley: The Rise and Fall of Media District* (New York: Routledge, 2004); Gina Neff, "Game Over," *American Prospect* 13 (2001): 16; and Andrew Ross, *No-Collar* (New York: Basic Books, 2002).
10. "Indie" is a slang derivative of "independent," part of the reaction in some pockets of 1990s youth culture against the presumably homogenized products of corporate cultural producers. Along with the label "alternative," it was intended to denote an

edgier, more experimental brand of cultural production, whether in film, fashion, or music. Many cultural critics, particularly the Chicago-based Tom Frank, have aggressively sought to expose these labels as highly misleading and fabricated poses that primarily benefit the very corporations they claim to oppose. See Thomas Frank and Matt Weiland, eds., *Commodify Your Dissent* (New York: Norton, 1997). For a treatment of the emerging indie scene of the 1980s, see Michael Azerrad, *Our Band Could Be Your Life* (New York: Back Bay Books, 2001).

11. Will Hogan, "West Town," in *The Local Community Fact Book of the Chicago Metropolitan Area* (Chicago: Chicago Review Press, 1984).

12. Tricia Drevets, "A Greening in Wicker Park," *Chicago Tribune*, Sept. 16, 1988, p. CN24.

13. Alan Artner, "Group Exhibits in 3 Wicker Park Galleries Worth a View," *Chicago Tribune*, Aug. 25, 1989, p. CN53; David McCracken, "Renier Charts New Territory in Wicker Park," *Chicago Tribune*, March 11, 1988, p. CN57.

14. William Julius Wilson has used the South Side as an exemplary platform from which to dissect the woes of the inner-city minority poor in the major works *The Truly Disadvantaged* (Chicago: University of Chicago Press, 1987) and *When Work Disappears* (New York: Knopf, 1996). See also Loic J. D. Wacquant, "Inside the Zone," in Pierre Bourdieu, *The Weight of the World* (Palo Alto, CA: Stanford University Press, 1999).

15. Boehlert, "Chicago: Cutting Edge's New Capital."

16. Michael Goldberg, "The Goose That Laid the Golden Eggs: Wicker Park Was Cutting Edge's Capital Last Year, but What About Now?" *Insider* 1994.

17. Jerry Shriver, "The Windy City's Burst of Bohemia," *USA Today*, Aug. 9, 2002.

18. Brenda Fowler, "The Many Accents of Wicker Park," *New York Times*, March 24, 2002.

19. Walter Benjamin, "Paris, Capital of the Nineteenth Century," in *The Arcades Project* (Cambridge, MA: Harvard University Press, 1999).

20. Quoted in Margy Rochlin, "Edgy in Chicago: The Music World Discovers Wicker Park," *New York Times*, March 14, 1993.

21. See Noel Riley Fitch, *Sylvia Beach and the Lost Generation: A History of Literary Paris in the Twenties and Thirties* (New York: Norton, 1985); Dan Franck, *Bohemian Paris* (New York: Grove, 2001); and Ernest Hemingway, *A Moveable Feast* (New York: Scribners, 1964).

22. Richard Florida, *The Rise of the Creative Class* (New York: Basic Books, 2002), p. 210.

23. Terry Nichols Clark, "Trees and Real Violins: Building Postindustrial Chicago," working paper, University of Chicago, n.d.

24. See Robert D. Atkinson, "Technological Change and Cities," *Cityscape* 3, 3 (1998): 129–170; Manuel Castells, *The Informational City* (Oxford: Blackwell, 1989); Florida, *The Rise of the Creative Class*; Saskia Sassen, *The Global City: New York, London, Tokyo* (Princeton, NJ: Princeton University Press, 2001).

25. See Michael Dear, *From Chicago to LA: Making Sense of Urban Theory* (Thousand Oaks, CA: Sage, 2002); Allen Scott and Edward Soja, eds., *The City: Los Angeles and Urban Theory at the End of the Twentieth Century* (Berkeley: University of California Press, 1996); Soja, *Postmodern Geographies: The Reassertion of Space in Critical Social Theory* (New York: Verso, 1989).

26. Saskia Sassen, "Cities: Between Local Actors and Global Conditions," *1997 Lefrak Monograph* (College Park, MD: Urban Studies and Planning Program, 1998); Neil Smith, *The New Urban Frontier* (New York: Routledge, 1996); Sharon Zukin, *The Culture of Cities* (London: Blackwell, 1995).

27. See Robert Park, Ernest Burgess, and Roderick McKenzie, *The City: Suggestions for Investigation of Human Behavior in the Urban Environment* (Chicago: University of Chicago Press, 1925). Roland Warren argues that "the city which contained the University of Chicago became the best researched city in the world" (*The Community in America* [New York: Rand McNally, 1973]). See also Andrew Abbott, *Department and Discipline: Chicago Sociology at One Hundred* (Chicago: University of Chicago Press, 1999); Robert Beauregard, "City of Superlatives," *City and Community* 2, 3 (2003): 183–199; and Lester Kurtz, *Evaluating Chicago Sociology* (Chicago: University of Chicago Press, 1984).

28. Michael Dear, *The Postmodern Urban Condition* (Oxford: Blackwell, 2000); Michael Dear, "Los Angeles and the Chicago School," *City and Community* 1, 1 (2002): 5–32.

29. Janet L. Abu-Lughod, *New York, Chicago, Los Angeles: America's Global Cities* (Minneapolis: University of Minnesota Press, 1999); William Testa, "A City Reinvents Itself," in *Global Chicago*, ed. Charles Madigan (Urbana and Chicago: University of Illinois Press).

30. Bourdieu, "The Abdication of the State," in *The Weight of the World*, p. 181.

31. Serge Guilbaut, *How New York Stole the Idea of Modern Art* (Chicago: University of Chicago Press, 1983).

32. Sharon Zukin, *Loft Living Culture and Capital in Urban Change* (Baltimore, MD: Johns Hopkins University Press, 1982), esp. pp. 176–182.

33. Rosalyn Deutsche, *Evictions: Art and Spatial Politics* (Cambridge: MIT Press, 1996); Deutsche and C. G. Ryan, "The Fine Art of Gentrification," October 31: 91–111, 1984; Christopher Mele, *Selling the Lower East Side* (Minneapolis: University of Minnesota Press, 2000); William Sites, *Remaking New York: Primitive Globalization and the Politics of Urban Community* (Minneapolis: University of Minnesota Press, 2003); Smith, *New Urban Frontier*.

34. Richard Florida, "Bohemia and Economic Geography," *Journal of Economic Geography* 2 (2002): 55–71; Ann Markusen and David King, "The Artistic Dividend: The Arts' Hidden Contribution to Regional Development," Project on Regional and Industrial Economics, Humphrey Institute of Public Affairs, 2003.

35. Cesar Grana, 1999.

36. Daniel Bell, *The Cultural Contradictions of Capitalism* (New York: Basic Books, 1976).

37. Florida, *Rise of the Creative Class*, p. 210.

Chapter 2

1. Barbara Vitello, "Wicker Park Says Its Farewell to Urbus Orbis," *Daily Herald*, Jan. 9, 1998.

2. Ned Polsky claims that the term "hip" is derived "directly from a much earlier phrase, 'to be on the hip,' to be a devotee of opium smoking — during which activity one lies

on one's hip. . . . Now the word has the generalized meaning of 'in the know' and even among beat drug users doesn't always refer specifically to knowledge of drugs" (*Hustlers, Beats, and Others* [New York: Anchor, 1967] pp. 145–146). John Leland proposes a different origin in *Hip: The History* (New York: HarperCollins, 2004), claiming it derives from the Wolof verb *hepi* ("to open one's eyes"); he too notes the term's migration from the fringes of society into more generalized and widely marketable usage.

3. Paul du Gay and Michael Pryke, eds., *Cultural Economy* (London: Sage, 2002); Angela McRobbie, *In the Culture Society: Art, Fashion, and Popular Music* (London: Routledge, 1999).

4. Fredric Jameson, *Postmodernism: or, the Cultural Logic of Late Capitalism* (Durham, NC: Duke University Press, 1991).

5. For example, Manuel Castells, *The Rise of the Network Society* (Oxford: Blackwell, 1996), pp. 407–459.

6. Harvey Molotch, *Where Stuff Comes From* (New York: Routledge, 2003); Molotch, "LA as Design Product," in *The City*, ed. Allen Scott and Edward Soja (Berkeley: University of California Press, 1996).

7. Henri Lefebvre, *The Production of Space* (Cambridge: Blackwell, 1974).

8. Urbus Orbis received substantial attention from both the local and the national media. In almost every case the adjective "bohemian" is used explicitly by the press to indicate what sort of place it is. June Sawyers, "A Land Without Boundaries," *Chicago Tribune*, Feb. 23, 1990, p. 7:3; Jae-Ha Kim, "Quiet Readers Digest Coffee and Culture at Urbus Orbis," *Chicago Sun-Times*, Aug. 2, 1991; Jodi Wilgoren, "A Place to Just Be Real," *Chicago Tribune*, Aug. 23, 1991, p. 7:3; Michael Gillis, "Cooler by the Expressway," *Chicago Sun-Times*, Oct. 23, 1992; Marcy Mason, "Outre Limits," *Chicago Tribune*, March 19, 1995, p. 5:1.

9. This mode of analysis is consistent with Michael Burawoy's extended case method, "extending from micro-processes," that is, local uses of the building space, "to macro forces," that is, the global economy. He further argues that "the macro-micro link refers not to . . . an expressive totality, but a structured one, in which the part is shaped by its relation to the whole, the whole being represented by external forces." Burawoy et al., *Global Ethnography* (Berkeley: University of California Press, 2000), p. 27.

10. Saskia Sassen, "Scales and Spaces: A Reply to Michael Dear," *City and Community* 1, 1 (2002): 49.

11. Rachel Weber, "Extracting Value from the City: Neoliberalism and Urban Redevelopment," *Antipode* 43, 3 (2002): 526.

12. Immigration rates in the United States declined significantly immediately after the Second World War, when European immigration was effectively choked off. Immigration has "rebounded smartly in recent decades, as a result of both less restrictive legal immigration and the entrance of undocumented workers," although "foreign-born residents [still] made up 10.4% of the US population in 2000 versus 13.4% in 1900" (Testa, "A City Reinvents Itself," p. 37). Contemporary immigration has stemmed population losses in large cities, with new entrants coming mainly from Latin America and Asia; it provides a particularly vulnerable labor force for sweatshop exploitation, especially in the case of undocumented workers.

13. Michael Piore, "The Economics of the Sweatshop," in *No Sweat,* ed. Andrew Ross (New York: Verso, 1997). There is evidence to suggest that sweatshop production is in fact returning as a feature of the global city, nourished by immigration from the developing world and abetted by reduced state oversight, declining labor organization, and the exceptional time sensitivity found in some fashion-intensive industries, especially women's apparel. Related chapters in *No Sweat:* Alan Howard, "Labor, History, and Sweatshops in the New Global Economy"; Steve Nutter, "The Structure and Growth of the Los Angeles Garment Industry"; and Julie Su, "El Monte Thai Garment Workers." See also Saskia Sassen, "The Informal Economy," in *Dual City: Restructuring New York,* ed. John Mollenkopf and Manuel Castells (New York: Russell Sage, 1991); and Edna Bonacich and Richard P. Appelbaum, "The Return of the Sweatshop," in *Cities and Society,* ed. Nancy Kleniewski (London: Blackwell, 2005).

14. See William Kornblum on the South Side steel mill economy in the late 1960s, *Blue Collar Community* (Chicago: University of Chicago Press, 1974).

15. Lefebvre, *Production of Space;* also Edward Soja, *Thirdspace: Journeys to Los Angeles and Other Real-Imagined Places* (Cambridge, MA: Blackwell, 1996).

16. See Jean Baudrillard, *For a Critique of the Political Economy of the Sign* (New York: Telos, 1981); Michael Hardt and Antonio Negri, *Empire* (Cambridge, MA: Harvard University Press, 2000); Scott Lash and John Urry, *Economies of Signs and Space* (Thousand Oaks, Calif.: Sage, 1994); and Maurizio Lazzarato, "Immaterial Labor" in *Marxism beyond Marxism,* ed. Saree Makdisi, Cesare Casarino and Rebecca E. Karl (London: Routledge, 1996).

17. Elaine A. Coorens, *Wicker Park From 1673 thru 1929* (Chicago: Old Wicker Park Committee, 2003), p. 10.

18. Ibid., pp. 43-48.

19. Abu-Lughod, *America's Global Cities,* p. 101.

20. Ross Miller, *The American Apocalypse: Chicagoans and the Great Fire, 1871–1874* (Chicago: University of Chicago Press, 1990).

21. As Ulf Hannerz notes, "This was a new city where no sentiments attached to particular areas had become strong enough to upset economic processes" — in contrast to the older European capitals. Hannerz, *Exploring the City* (New York: Columbia University Press, 1980), p. 28. Soja considers Chicago and Manchester, England, the exemplars of what he calls "the second urban revolution," meaning industrial urbanism, in *Postmetropolis: Critical Studies of Cities and Regions* (Cambridge, UK: Blackwell, 2000).

22. W. I. Thomas and Florian Znaniecki, *The Polish Peasant in Europe and America* (New York: Knopf, 1927); Melvin Holli and Peter d'A. Jones, eds., *Ethnic Chicago* (Grand Rapids, MI: Eerdmans, 1977).

23. James Grossman, "African American Migration to Chicago," in *Ethnic Chicago,* ed. Melvin Holli and Peter d'A. Jones (Grand Rapids, MI: Eerdmans, 1977).

24. Harvey Warren Zorbaugh, *The Gold Coast and the Slum* (Chicago: University of Chicago Press, 1929), p. 1.

25. Robert E. Park, "The City: Suggestions for the Investigation of Human Behavior in the Urban Environment," in Robert E. Park, Ernest W. Burgess, and

Roderick D. McKenzie, *The City: Suggestions for Investigation of Human Behavior in the Urban Environment* (Chicago: University of Chicago Press, 1925), pp. 10, 40.

26. On Burgess' map, Wicker Park is subsumed by a larger community area known as West Town. I have never heard this term used by a local resident. However, Burgess' map is commonly used to subaggregate data, since its durability allows for valid comparisons over time. The census data I have compiled corresponds to the generally agreed upon boundaries of Wicker Park, not West Town. However, I also refer to studies that encompass the larger area, and readers should be alert to this fact. Despite Burgess' oversight, Albert Hunter (*Symbolic Communities* [Chicago: University of Chicago Press, 1974]) and Coorens (*Wicker Park from 1673*) both indicate that Wicker Park was recognized as a distinct community area by residents at least by the early twentieth century.

27. Coorens, *Wicker Park from 1673*, p. 80.

28. Ibid., pp. 80–81; Hogan, "West Town," 1984.

29. Dominic A. Pacyga and Ellen Skerrett, *Chicago: City of Neighborhoods* (Chicago: Loyola University Press, 1986), p. 172.

30. Coorens, *Wicker Park from 1673*, pp. 44–45.

31. Dave Hoekstra, "Dancing Down 'Polish Broadway,'" *Chicago Sun-Times*, Feb. 28, 1994, p. 2:24; Pacyga and Skerret, *Chicago*, p. 165.

32. Edward Kantowicz, "Polish Chicago," in *Ethnic Chicago,* ed. Melvin Holli and Peter d'A. Jones (Grand Rapids, MI: Eerdmans, 1977), p. 175.

33. Ibid.

34. Helena Lopata, "The Function of Voluntary Associations in an Ethnic Community: 'Polonia,'" in *Contributions to Urban Sociology,* ed. Ernest W. Burgess and Donald J. Bogue (Chicago: University of Chicago Press, 1964).

35. Hunter, *Symbolic Communities*, p. 25.

36. Gerald Suttles, William Kornblum, and Albert Hunter are prominent scholars that carried on the Chicago School approach to urban studies into the 1970s and beyond.

37. Manuel Castells, *The Urban Question* (Cambridge, MA: MIT Press, 1972); Mark Gottdiener, *The Social Production of Urban Space* (Austin: University of Texas Press, 1985); Hannerz, *Exploring the City*; David Harvey, "On Countering the Marxian Myth — Chicago Style," in *Spaces of Capital* (New York: Routledge, 2001).

38. Greg Hise, "Industry and the Landscapes of Social Reform," in *From Chicago to LA,* ed. Michael Dear (Thousand Oaks, CA: Sage, 2002), p. 99.

39. George Ritzer, *The McDonaldization of Society* (Thousand Oaks, CA: Pine Forge Press, 1996).

40. Adam Smith made this point quite vividly more than a century earlier in his analysis of the pin factory. See Smith, *The Wealth of Nations* (New York: Bantam [1793] 2003), pp. 4–5.

41. Antonio Gramsci, "Americanism and Fordism," in *Selections from the Prison Notebooks* (New York: International, 1971), p. 286.

42. Michel Aglietta, *A Theory of Capitalist Regulation: The U.S. Experience* (New York: Verso, 1979).

43. Friedrich Engels, *The Condition of the Working-Class in England in 1844* (London: Swan Sonnenschein, 1892).

44. Louis Wirth, *The Ghetto* (Chicago: University of Chicago Press, 1928); Zorbaugh, *Gold Coast*; also see Jane Addams, *Twenty Years at Hull House* (New York: McMillan, 1912).

45. Ira Katznelson, *City Trenches* (Chicago: University of Chicago Press, 1982); Gerald Suttles, *The Social Construction of Community* (Chicago: University of Chicago Press, 1973).

46. Aglietta, *Theory of Capitalist Regulation*; Robert Boyer, *The Regulation School: A Critical Introduction* (New York: Columbia University Press, 1990); Alain Lipietz, *The Enchanted World* (London: Verso, 1985).

47. Abu-Lughod, *America's Global Cities,* p. 217.

48. Joseph Bensman and Arthur J. Vidich, *American Society: The Welfare State and Beyond* (South Hadley, MA: Bergin and Garvey, 1987).

49. Castells, *The Informational City* (Oxford: Blackwell, 1989), p. 21.

50. John Kasarda, "Urban Change and Minority Opportunities," in *The New Urban Reality,* ed. P. E. Peterson (Washington, D.C.: Brookings Institution, 1985).

51. Amos Hawley, *Human Ecology* (New York: Ronald, 1950); Suttles, *Social Construction of Community.*

52. Kornblum, *Blue-Collar Community.*

53. Ibid., p. 18.

54. Bensman and Vidich, *American Society,* p. 313.

55. Daniel Bell, *The End of Ideology* (Cambridge, Mass.: Harvard University Press, 1960); John Kenneth Galbraith, *The Affluent Society* (New York: Mariner, 1998).

56. Joel Krieger, *Reagan, Thatcher, and the Politics of Decline* (Oxford: Oxford University Press, 1986).

57. Robert Beauregard, *Voices of Decline: The Fate of Postwar Cities* (New York: Routledge, 2003); Terry Nichols Clark and Lorna Ferguson, *City Money* (New York: Columbia University Press, 1983).

58. David Harvey, *The Condition of Postmodernity* (Cambridge UK: Blackwell, 1989), p. 142.

59. Ibid., pp. 141–172.

60. Adam Cohen and Elizabeth Taylor, *American Pharaoh: Mayor Richard J. Daley: His Battle for Chicago and the Nation* (New York: Little, Brown, 2000); Mike Royko, *Boss: Richard J. Daley of Chicago* (New York: Dutton, 1971).

61. Beauregard, *Voices of Decline.*

62. Hunter, *Symbolic Communities,* p. 46.

63. Herbert Gans, "Urbanism and Suburbanism as Ways of Life," in *Metropolis: Center and Symbol of Our Times,* ed. Philip Kasinitz (New York: New York University Press, 1995); Kenneth Jackson, *Crabgrass Frontier: The Suburbanization of the United States* (Cambridge: Oxford University Press, 1985).

64. Kasarda, "Urban Change and Minority Opportunities."

65. Kornblum, *Blue Collar Community*; Katznelson, *City Trenches.*

66. Annette Fuentes and Barbara Ehrenreich, *Women in the Global Factory* (Cambridge, MA: South End Press, 1984); William Finnegan, "The Economics of Empire," *Harper's,* May 2003; Naomi Klein, *No Logo: Money, Marketing, and the Growing Anti-Corporate Movement* (New York: Picador, 1999).

67. Margaret Villanueva, Brian Erdman and Larry Howlett, "World City/Regional City: Latinos and African Americans in Chicago and St. Louis," working paper no. 46, Julian Samora Research Institute, Michigan State University, 2000.
68. Thomas W. Lester, "Old Economy or New Economy? Economic and Social Change in Chicago's West Town Community Area," master's thesis, Urban Planning and Policy department, University of Illinois–Chicago, 2000.
69. Hogan, "West Town."
70. Jeff Huebner, "The Panic in Wicker Park," *Chicago Reader*, Aug. 26, 1994.
71. Louis Wirth, "Urbanism as a Way of Life," *American Journal of Sociology* 44 (1938): 3–24.
72. Saskia Sassen, *Cities in a World Economy* (Thousand Oaks, CA: Pine Forge Press, 1994), p. 2.
73. Gottdiener, *Social Production of Urban Space.*
74. Joel Garreau, *Edge City: Life on the New Frontier* (New York: Anchor, 1991).
75. Dear, "Los Angeles and the Chicago School," p. 6.
76. Sassen, *Cities in a World Economy,* p. 2.
77. Sassen, *Global City.*
78. Ibid., p. 6.
79. William Ferris, *The Grain Traders: The Story of the Chicago Board of Trade* (East Lansing: Michigan State University Press, 1988).
80. Terry Nichols Clark and Michael Rempel, *Citizen Politics in Postindustrial Societies* (Boulder, CO: Westview Press, 1997); Terry Nichols Clark et al., "Amenities Drive Urban Growth," *Journal of Urban Affairs* 24, 5 (2002): 517–532; Florida, *Rise of the Creative Class.*
81. See John Mollenkopf and Manuel Castells, eds., *Dual City: Restructuring New York* (New York: Russell Sage Foundation, 1991).
82. Harvey, *Condition of Postmodernity,* p. 156.
83. Klein, *No Logo.*
84. William Lilley and Laurence DeFranco, "Geographic and Political Distribution of Arts-Related Jobs in Illinois," study commissioned by the Illinois Arts Alliance Foundation, 1995, p. 7.
85. Sharon Zukin's study of the redevelopment of SoHo, *Loft Living,* while primarily concerned with the real estate market, also posits the role of the district in organizing the "artistic mode of production" (p. 176). Similar themes have been addressed more recently in Florida's *Rise of the Creative Class* and Joel Kotkin's "Here Comes the Neighborhood," *inc.com,* July 1, 2000.
86. William Julius Wilson, *Truly Disadvantaged* and *When Work Disappears.*
87. George Gilder, *Wealth and Poverty* (New York: Basic Books, 1981); Charles Murray, *Losing Ground* (New York: Basic Books, 1984); James Q. Wilson, *Thinking About Crime* (New York: Basic Books, 1975).
88. Deutsche and Ryan, "The Fine Art of Gentrification;" Mele, *Selling the Lower East Side*; Neil Smith, *New Urban Frontier.*

89. Smith, *New Urban Frontier*, pp. 7–8.

90. Ibid., p. 68.

Chapter 3

1. Ross Wetzsteon, *Republic of Dreams: Greenwich Village, the American Bohemia, 1910–1960* (New York: Simon and Schuster, 2002), p. 11.

2. Marshall Berman, *All That Is Solid Melts Into Air* (New York: Penguin, 1982), p. 16.

3. Ibid., p. 15; Karl Marx and Friedrich Engels, *The Communist Manifesto* (New York: Signet, 1998).

4. Berman, *All That Is Solid*, p. 16.

5. Jerrold Seigel, *Bohemian Paris* (Baltimore: Johns Hopkins University Press, 1986); Christine Stansell, *American Moderns: Bohemian New York and the Creation of a New Century* (New York: Owl, 2000).

6. George Levitine, *The Dawn of Bohemianism* (University Park: Pennsylvania State University Press, 1978).

7. Mary Gluck, *Popular Bohemia: Modernism and Urban Culture in Nineteenth Century Paris* (Cambridge, MA: Harvard University Press, 2005).

8. George Becker, *The Mad Genius Controversy* (Beverly Hills, CA: Sage, 1978).

9. Marcus Boon, *The Road of Excess: A History of Artists on Drugs* (Cambridge, MA: Harvard University Press, 2002).

10. Berman, *All That Is Solid*, pp. 155–164.

11. Andrea Barnet, *All-Night Party: The Women of Bohemian Greenwich Village and Harlem* (Chapel Hill, NC: Algonquin, 2004); Jean Moorcraft Wilson, *Virginia Woolf's London* (New York: Tauris Park, 2000); Ulf Hannerz, *Cultural Complexity* (New York: Columbia University Press, 1992); Steven Watson, *The Harlem Renaissance* (New York: Pantheon, 1995).

12. Walter Benjamin, "Paris."

13. George Snyderman and William Josephs, "Bohemia: The Underworld of Art," *Social Forces* 18, 2 (1939): 187.

14. Ibid.

15. Cesar Grana, *Bohemian vs. Bourgeois* (New York: Basic Books, 1964).

16. Malcolm Easton, "Literary Beginnings," and V. S. Pritchet, "Murger's *La Vie de Bohème*," in *On Bohemia*, ed. Cesar Grana and Marigold Grana (New Brunswick, NJ: Transaction, 1990); Seigel, *Bohemian Paris*, pp. 31–38.

17. The opera opened in Turin in 1896; its New York premiere (at the Met) was in 1900. Giacomo Puccini, *La Bohème* (New York: Riverrun, 1982).

18. Grana, *Bohemian vs. Bourgeois*; Pierre Bourdieu, *The Rules of Art* (Palo Alto, CA: Stanford University Press, 1992).

19. Seigel, *Bohemian Paris*, p. 13.

20. Grana, "The Ideological Significance of Bohemian Life," in *On Bohemia*, ed. Cesar Grana and Marigold Grana (New Brunswick, NJ: Transaction, 1990), p. 4.

21. Grana, *Bohemian vs. Bourgeois*, p. 25.

22. Pierre Bourdieu, *The Field of Cultural Production* (New York: Columbia University Press, 1993).

23. Bourdieu, *Rules of Art*; David Harvey, *Paris: Capital of Modernity* (New York: Routledge, 2004).

24. Berman, *All That Is Solid*, pp. 134–142.

25. T. J. Clark, *The Painting of Modern Life* (Princeton, NJ: Princeton University Press, 1984); Cesar Grana, "Impressionism as an Urban Art Form," in *Fact and Symbol: Essays on the Sociology of Art and Literature* (London: Transaction, 1994).

26. T. J. Clark, *The Absolute Bourgeois* (Berkeley: University of California Press, 1973).

27. Bourdieu, *Rules of Art*; Seigel, *Bohemian Paris*.

28. On this distinction, see Pierre Bourdieu, *Outline of a Theory of Practice* (Cambridge: Cambridge University Press, 1977); William H. Sewell Jr., "A Theory of Structure: Duality, Agency and Transformation," *American Journal of Sociology* 98 (1992): 1-29.

29. Seigel, *Bohemian Paris*, p. 4.

30. Grana, *Bohemian vs. Bourgeois*, p. 23.

31. Grana, "The Ideological Significance of Bohemian Life," p. 3.

32. Bourdieu, *Field of Cultural Production*.

33. Ephraim Mizruchi, "Bohemia as a Means of Social Regulation," in *On Bohemia*, ed. Cesar Grana and Marigold Grana (New Brunswick, NJ: Transaction, 1990), p. 14.

34. George Becker, *Mad Genius Controversy*.

35. David Brooks, *BoBos in Paradise* (New York: Simon and Schuster, 2001), p. 67. Brooks means by "us," educated readers who, he presumes, are not only familiar with the lifestyle of bohemia but incorporate elements of it into their own performative strategies — a claim we will take up later.

36. Stansell, *American Moderns*, p. 17.

37. Ibid., p. 43.

38. Caroline F. Ware, *Greenwich Village: 1920–1930* (New York: Harper Colophon, 1935), p. 3.

39. Max Eastman, *Art and the Life of Action* (New York: Books for Libraries, 1970); William O'Neil, *The Last Romantic: A Life of Max Eastman* (Oxford: Oxford University Press, 1978).

40. Stansell, *American Moderns*, p. 12.

41. Zorbaugh, *Gold Coast*, p. 7.

42. Ibid., p. 87.

43. Ibid., p. 102.

44. Stansell, *American Moderns*, p. 4.

45. Alson Smith, *Chicago's Left Bank* (Chicago: Henry Regnery, 1953).

46. Zorbaugh, *Gold Coast*, p. 91.

47. Louis Wirth, "Urbanism as a Way of Life."

48. Theodor W. Adorno and Max Horkheimer, *Dialectic of Enlightenment* (New York: Continuum, 1994), p. 121.

49. Ibid., p. 120.

50. John Gruen, *The New Bohemia* (New York: Grosset and Dunlap, 1966); Polsky, *Hustlers*.

51. Lawrence Lipton, *The Holy Barbarians* (New York: Grove Press, 1959).

52. Hannerz, *Cultural Complexity.*

53. Ned Polsky, "Reflections on the Hip," in Norman Mailer, *Advertisements for Myself* (New York: Putnam, 1959).

54. Jack Kerouac, *On the Road* (New York: Penguin, 1955).

55. Ibid., p. 180.

56. Norman Mailer, "The White Negro," in *Advertisements for Myself* (New York: Putnam, 1959), p. 314.

57. Ibid.

58. Bettina Drew, *Nelson Algren: A Life on the Wild Side* (Austin: University of Texas Press, 1989).

59. Herbert Marcuse, *One-Dimensional Man* (Boston: Beacon, 1964), p. 256.

60. Daniel Bell, "New Afterword" to *The End of Ideology* (Cambridge, MA: Harvard University Press, 1988).

61. David McBride, "Death City Radicals: The Counterculture in the New Left in 1960's Los Angeles," in *The New Left Revisited,* ed. John McMillian and Paul Buhle, (Philadelphia: Temple University Press, 2002).

62. Tom Wolfe, *The Electric Kool-Aid Acid Test* (New York: Bantam, 1968).

63. Allan Bloom, *The Closing of the American Mind* (New York: Simon and Schuster, 1987); Fredric Jameson, "Periodizing the 1960's," in *The Sixties Without Apology,* ed. Sohnya Sayres et al. (Minneapolis: Minnesota University Press, 1984); Bell, *Cultural Contradictions of Capitalism.*

64. Bell, *Cultural Contradictions of Capitalism,* p. 40.

65. Todd Gitlin, *The Sixties: Years of Hope, Days of Rage* (New York: Bantam, 1987), pp. 196–197.

66. Harvey, *Condition of Postmodernity.*

67. Perry Anderson, *The Origins of Postmodernity* (New York: Verso, 1998), p. 55.

68. Bell, *Cultural Contradictions of Capitalism,* p. 41.

69. Ann Markusen, "Targeting Occupations Rather Than Industries in Regional and Economic Development," paper presented at the North American Regional Science Association Meetings, Chicago, 2000.

70. Brooks, *BoBos.*

71. Florida, *Rise of the Creative Class,* pp. 190–211.

72. Bell, *Cultural Contradictions of Capitalism.*

73. Thomas Frank, *The Conquest of Cool* (Chicago: University of Chicago Press, 1997).

74. Harvey, *Condition of Postmodernity;* Fredric Jameson, *The Cultural Turn: Selected Writings on Postmodernism* (New York: Verso, 1998)..

75. Florida, *Rise of the Creative Class.*

76. Ibid., p. 61.

77. Ibid., p. 210.

78. Ibid.

79. At one point Florida insists that Marx's categories of the bourgeoisie and the proletariat no longer make much sense, since the "means of production" now reside in the heads of the creative class.

80. See Thomas Frank's extensive critique of new economy rhetoric, *One Market Under God* (New York: Anchor, 2000).

81. Florida, *Rise of the Creative Class*, p. 177.

Chapter 4

1. John Kass, "Two Charged in Gang Killing," *Chicago Tribune*, Oct. 27, 1986, p. C2.

2. James David Nathan, "Wicker Park's Constant Factor: Change," *Chicago Tribune*, Oct. 1, 1989.

3. Neil Smith, *New Urban Frontier*.

4. A. Langer and Mark Rattin, "Neighborhood on the Verge," *Chicago SubNation*, 1994 11:2, p. 17.

5. Mele, *Selling the Lower East Side*, p. 233.

6. Christopher Mele, "The Process of Gentrification in Alphabet City," in *From Urban Village to East Village*, ed. Janet L. Abu-Lughod (Oxford, UK: Blackwell, 1994), p. 186.

7. Reymundo Sanchez's memoir detailing his life as a Latin King takes place around Wicker Park and the neighboring district of Humboldt Park in the mid-1980s. Sanchez, *My Bloody Life* (Chicago: Chicago Review Press, 2000).

8. See Elijah Anderson, *Code of the Street: Decency, Violence, and the Moral Life of the Inner City* (New York: Norton, 1999), for a discussion of the relationship between respectability and criminality in a low-income urban community.

9. Simple loitering in low-income communities has been increasingly construed as a sign of disorder and, according to the "broken windows" model of community control, is therefore likely to contribute to more serious criminal activities. See Robert Jackal, "What Kind of Order?" *Criminal Justice Ethics* (summer 2003): 54–67; Robert J. Sampson and Steven Raudenbush, "Seeing Disorder: Neighborhood Stigma in the Social Construction of 'Broken Windows'" *Social Psychology Quarterly* 67 (2003): 319–342.

10. Elijah Anderson, *Streetwise: Race, Class, and Change in an Urban Community* (Chicago: University of Chicago Press, 1990).

11. See Malcolm Gladwell, "The Coolhunt," and Alex Kotlowicz, "False Connections," both in *The Consumer Society Reader*, ed. Juliet B. Schor and Douglas B. Holt (New York: New Press, 2002).

12. See Angela McRobbie's critique of the masculinist assumptions of subcultural studies, "Settling Accounts with Subcultures: A Feminist Critique," *Screen Education* 34 (1980): 37–49.

13. Great rivers of ink were spent in their epistolary romance, largely collected in Simone de Beauvoir, *Beloved Chicago Man: Letters to Nelson Algren 1947–64* (London: Gollancz, 1998).

14. Holden, *Literary Chicago*.

15. Nelson Algren, *The Man with the Golden Arm* (New York: Seven Stories, 1949).

16. Carlo Rotella, *October Cities* (Berkeley: University of California Press, 1998), p. 10.

17. Ibid., p. 3.

18. Nelson Algren, "The Devil Came Down Division Street," in *The Neon Wilderness* (New York: Doubleday, 1947).

19. Art Shay, *Nelson Algren's Chicago* (Urbana-Champaign: University of Illinois Press, 1988).

20. Nelson Algren, *Never Come Morning* (New York: Seven Stories, 1942).

21. Tama Janowitz, *Slaves of New York* (New York: Washington Square, 1986) p. 3.

22. John Blades, "Just Boggling: Wicker Park Bookstore Owner Wonders Who's Behind Shop's Phantom Boycott," *Chicago Tribune*, July 29, 1993, p. C11.

23. Klein, *No Logo*.

24. Honore de Balzac, *Lost Illusions* (New York: Viking, [1843] 1976); Gustave Flaubert, *Sentimental Education* (New York: Viking, [1869] 1991).

25. Regenia Gagnier, *The Insatiability of Human Wants* (Chicago: University of Chicago Press, 2000), p. 210.

26. Mitchell Duneier, *Sidewalk* (New York: Farrar, Strauss, and Giroux, 1999).

27. Jane Jacobs, *The Death and Life of Great American Cities* (New York: Basic Books, 1961).

28. Jim DeRogatis, "Outsider Influence: Singer's Raves Spur Exploitation Debate," *Chicago Sun-Times*, 1995, p. NC5.

29. Maxwell Bodenheim, *My Life and Loves in Greenwich Village* (New York: Bridgehead Press, 1954).

30. DeRogatis, "Outsider Influences," p. 4.

31. Phoebe Hoban, *Basquiat: A Quick Killing in Art* (New York: Penguin, 1998).

32. Joseph Mitchell, *Up in the Old Hotel* (New York: Vintage, 1993).

33. Wetzsteon, *Republic of Dreams*, p. 419.

34. Buddy Seigal, "Willis: Both Disturbed, Disturbing," *Los Angeles Times*, March 16, 1996, p. F2.

35. Libby Copeland, "Songs in His Head: Wesley Willis Is Haunted by Voices. His Music Helps to Drown Them Out," *Washington Post*, Nov. 24, 2000, p. C1.

36. Ibid.

37. Mike Davis, *City of Quartz* (New York: Verso, 1990). See also Madeleine Stoner, "The Globalization of Urban Homelessness," in *From Chicago to L.A.*, ed. Michael Dear (Thousand Oaks, CA: Sage, 2002).

38. John Pareles, "When the Suffering Undoes the Artist," *New York Times*, April 28, 2002, 2:1.

39. William S. Burroughs, *Junky* (New York: Ace, 1953); Burroughs, *Naked Lunch* (Paris: Olympia, 1959); Jim Carroll, *The Basketball Diaries* (Bolinas, CA: Tombouctou, 1978).

40. Quoted in Katie Schoemer, "Rockers, Models, and the New Allure of Heroin," *Newsweek*, Aug. 26, 1996.

41. Andreas Glaeser, *Divided in Unity* (Chicago: University of Chicago Press, 2000), p. 9.

42. Herbert Gans, *Popular Culture and High Culture* (New York: Basic Books, 1974).

43. Jeff Huebner, "The Panic in Wicker Park," *Chicago Reader*, Aug. 26, 1994.

44. Renato Rosaldo, *Culture and Truth* (Boston: Beacon, 1993), p. 69.

45. John Irwin, *Scenes* (London: Sage, 1977).

46. Brenda Fowler, "The Many Accents of Wicker Park," *New York Times*, March 24, 2002, Travel section, p. 1.

Chapter 5

1. Ray Oldenberg, *The Great Good Place* (New York: Paragon, 1989).

2. David McCracken, "Reneir Charts New Territory in Wicker Park," *Chicago Tribune*, March 11, 1988, p. CN57.

3. Schippers, *Rockin' Out of the Box*, p. 42.

4. Neil Smith illustrates how the imagery of the frontier wilderness was incorporated into the marketing of the East Village during the 1980s, encouraging affluent residents to identify with the romantic imagery of the Old West pioneer. See Neil Smith, "New City, New Frontier: The Lower East Side as Wild, Wild West," in *Variations on a Theme Park*, ed. Michael Sorkin (New York: Hill and Wang, 1992).

5. Huebner, "Panic in Wicker Park."

6. Lewis Lazare, "The Coyote's Latest Howl," *Chicago Reader*, April 21, 1995.

7. Seigel, *Bohemian Paris*, p. 37.

8. Kathryn Shattuck, "Kerouac's Road Scroll Is Going to Auction," *New York Times*, March 22, 2001.

9. Herbert Gold, *Bohemia: Where Love, Angst and Strong Coffee Meet* (New York: Simon and Schuster, 1993).

10. Marcy Mason, "Outre Limits," *Chicago Tribune*, March 19, 1995, p. 1.

11. Jae-Ha Kim, "The House Blend: Urbus Orbis Draws Mix of Regulars," *Chicago Sun Times*, Dec. 26, 1993, p. 7.

12. Gerald Suttles, *The Man-Made City* (Chicago: University of Chicago Press, 1990).

13. Herbert J. Gans, *The Urban Villagers* (New York: Free Press, 1962).

14. Barbara Vitello, "Wicker Park Says Its Farewell to Urbus Orbis," (Arlington Heights) *Daily Herald*, Jan. 9, 1998.

15. J.H.K., "Guyville Java Jive," *Rolling Stone*, Aug. 11, 1994, p. 22.

16. Jancee Dunn, "Liz Phair," *Rolling Stone*, Oct. 6, 1994.

17. Wetzsteon, *Republic of Dreams*, p. 7

18. Zukin, *Loft Living*, p. 82.

19. Serge Guilbaut, *How New York Stole the Idea of Modern Art.*

20. Thomas Crow, *Modern Art and the Common Culture* (New Haven, CT: Yale University Press, 1996).

21. Richard Hofstadter, *Anti-Intellectualism in American Life* (New York: Random House, 1966).

22. Andrew Ross, *No Respect: Intellectuals and Popular Culture* (New York: Routledge, 1989). Even recent films with wildly gifted geniuses as their heroes, such as Gus Van Sant's *Good Will Hunting* and *Finding Forrester*, seem to hate intellectuals as a class.

23. Carole Vance, "The War on Culture," *Art in America* 47 (1989): 39–43.

24. Robert Reich, *The Work of Nations: Preparing Ourselves for Twenty-First-Century Capitalism* (New York: Knopf, 1991).

25. Florida, *Rise of the Creative Class*.

26. Pierre Bourdieu, *In Other Words* (London: Polity, 1990), p. 132.

27. Sarah Thornton, *Club Cultures: Music, Media, and Subcultural Capital* (Hanover, NH: University Press of New England, 1996), p. 99.

28. M. Brodsky, "Labor Market Flexibility," *Monthly Labor Review* 117, 11 (1994): 53–60; Paul Hirsch and Mark Shanley, "The Rhetoric of Boundaryless: How the Newly Empowered and Fully Networked Managerial Class of Professionals Bought Into and Self-Managed Its Own Marginalization," in *Boundaryless Careers*, ed. Michael Arthur and Denise Rousseau (Oxford: Oxford University Press, 1996); Katherine Newman, *Falling from Grace: The Experience of Downward Mobility in the American Middle Class* (New York: Vintage, 1988); Vicki Smith, "New Forms of Work Organization," *Annual Review of Sociology* 23 (1997): 315–339.

29. Suttles, *Man-Made City*, p. 97.

30. Smith, *New Urban Frontier*, p. 92.

31. Richard Lloyd and Terry Nichols Clark, "The City as an Entertainment Machine," *Research in Urban Sociology* 6 (2001): 359–380.

Chapter 6

1. Lloyd and Clark, "City as an Entertainment Machine."

2. Terry Nichols Clark, "Trees and Real Violins: Building Postindustrial Chicago," working paper, University of Chicago, 2000, p. 12.

3. Bart Eeckhout, "The Disneyfication of Times Square: *Back to the Future*," *Critical Perspectives on Urban Redevelopment* 6 (2001): 379–428; John Hannigan, *Fantasy City: Pleasure and Profit in the Postmodern Metropolis* (New York: Routledge, 1998).

4. Dennis Judd, "Constructing the Tourist Bubble," in *The Tourist City*, ed. Dennis Judd and Susan Fainstein (New Haven, CT: Yale University Press, 1999).

5. Jameson, *Postmodernism*, pp. 39–44; for a critique of Jameson's deracinated perspective, see Robert Fairbanks, "A Theoretical Primer on Space," *Critical Social Work* 3 (2003): 131–154.

6. Jean Baudrillard, *America* (New York: Verso, 1989); Umberto Eco, *Travels in Hyperreality* (New York: Harcourt Brace Jovanovich, 1986).

7. Michael Sorkin, *Variations on a Theme Park* (New York: Hill and Wang, 1992), p. 205.

8. Adorno and Horkheimer, *Dialectic of Enlightenment*.

9. Jean-François Lyotard, *The Postmodern Condition* (Minneapolis: University of Minnesota Press, 1979), p. 76.

10. David Grazian, *Blue Chicago: The Search for Authenticity in Urban Blues Clubs* (Chicago: University of Chicago Press, 2003).

11. Lloyd and Clark, "City as an Entertainment Machine," p. 357.

12. Sassen, *Global City*, pp. 286–287.

13. Clark and Rempel, *Citizen Politics in Postindustrial Societies*; Lloyd and Clark, "City as an Entertainment Machine."

14. Clark et al., "Amenities Drive Urban Growth"; Leonard Nevarez, *New Money, Nice Town* (New York: Routledge, 2003).

15. Richard A. Peterson and Roger Kern, "Changing Highbrow Taste: From Snob to Omnivore," *American Sociological Review* 61 (1996): 5.

16. Lloyd and Clark, "The City as an Entertainment Machine"

17. See James Naremore, *More Than Night: Film Noir in Its Contexts* (Berkeley: University of California Press, 1998).

18. Zukin, *Culture of Cities*, p. 154.

19. Ibid., p. 182.

20. Jacquee Thomas, "Transitional Areas Attracting Renters," *Chicago Sun-Times*, Sept. 20, 1998, p. NC2.

21. Jacobs, *Death and Life of Great American Cities*.

22. Erik Olin Wright, *Classes* (New York: Verso, 1985).

23. David Grazian makes a similar observation about the boorishness of dabblers in Chicago Blues clubs. Grazian, *Blue Chicago*.

24. Chicago designer Suhail, described in the *Chicago Tribune Magazine* as "one of the city's most creative and edgy design Turks," was responsible for rehabilitating and decorating the loft. The furniture was assembled almost entirely from retailers in the Wicker Park and Bucktown neighborhoods, and the walls were hung with the works of Chicago artists. I had the opportunity to walk through the loft before they dismantled it after the cast had left. Though the furnishings were chic, they were not very functional, a fact the cast members conceded to the press as well. See Lisa Skolnik, "Made for TV," *Chicago Tribune Magazine*, Jan. 13, 2002.

25. Alison Pollet, *MTV's Real World Chicago* (New York: MTV Books, 2002).

26. Lee Wilson Bailey, "For Reality Stars, the Reality of a One-Time Payment," *New York Times*, Dec. 16, 2001.

27. Bill Carter, "Reality Island: Overcrowded, but Showing No Signs of Sinking," *New York Times*, July 17, 2001.

28. Bailey, "For Reality Stars."

29. Michael Joseph Gross, "The Real World Gets a Script and Gets Unreal," *New York Times*, Aug. 4, 2002.

30. The twenty-four contestants on ABC's *Survivor* compete for a single prize of $1 million. If we assume that every contestant has an equal chance of winning (and the winner of the first installment, Richard Hatch, sure seemed unlikely at the outset), then the expected value of participation is $1 million divided by 24, or $41,667, a bonanza by reality standards. But this is a classic example of Robert H. Frank's concept of a winner-take-all market.

31. Mark Brown, "Unwanted Attention a Real Pain for MTV," *Chicago Sun-Times*, Aug. 11, 2001.

32. Kim, "Griping, Groping on North Avenue" *Chicago Sun Times*, Aug. 16, 2001.

33. William Rice, "A Piece of the Pie," *Chicago Tribune Magazine*, Jan. 13, 2002.

Chapter 7

1. Nelson Algren, *Chicago: City on the Make* (Chicago: University of Chicago Press, 1951), p. 23.
2. Crow, *Modern Art and the Common Culture*, p. 34.
3. Adorno and Horkheimer, *Dialectic of Enlightenment.*
4. Michael Piore and Charles F. Sabel, *The Second Industrial Divide* (New York: Basic Books, 1984); Michael Storper, "The Transition to Flexible Specialization in the US Film Industry: *External Economies, the Division of Labor, and the Crossing of Industry Divides," Cambridge Journal of Economics* 13 (1989): 273–305.
5. Scott Lash and John Urry, *Economies of Signs and Space* (Thousand Oaks, CA: Sage, 1994).
6. Joshua Gamson, *Claims to Fame: Celebrity in Contemporary America* (Berkeley: University of California Press, 1994), p. 25; Thomas Schatz, *The Genius of the System: Hollywood Filmmaking in the Studio Era* (New York: Owl, 1996).
7. William T. Bielby and Denise D. Bielby, "Organizational Mediation of Project-Based Labor Markets: Talent Agencies and the Careers of Screenwriters," *American Sociological Review* 64 (1999): 64–85.
8. Richard A. Peterson and N. Anand, "How Chaotic Careers Create Orderly Fields," in *Career Creativity: Explorations in the Remaking of Work*, ed. Maury Peiperl et al. (New York: Oxford University Press, 2002).
9. Seigel, *Bohemian Paris.*
10. Richard Caves, *Creative Industries: Contacts Between Art and Commerce* (Cambridge, MA: Harvard University Press, 2000), p. 75.
11. Pierre-Michel Menger, "Artistic Labor Markets and Careers," *Annual Review of Sociology* 25 (1999): 556.
12. Robert H. Frank and Phillip J. Cook, *The Winner-Take-All Society* (New York: Penguin, 1995).
13. Caves, *Creative Industries,* pp. 21–36; Randall K. Filer, "The Starving Artist: Myth or Reality? Earnings of Artists in the United States" *Journal of Political Economy* 94, 1 (1986): 56–75.
14. Mizruchi, "Bohemia as a Means of Social Regulation."
15. Edward O. Laumann et al., *The Sexual Organization of the City* (Chicago: University of Chicago Press, 2004).
16. Filer, "Starving Artist," p. 63.
17. Menger, "Artistic Labor Markets."
18. Vance, "War on Culture."
19. Stuart Plattner, *High Art Down Low* (Chicago: University of Chicago Press, 1996), p. 39.
20. Chin-tao Wu, *Privatising Culture: Corporate Art Intervention Since the 1980s* (New York: Verso, 2002).
21. In an intriguing interpretation of the economics of artistic production, Hans Abbing notes that in addition to exposure to the market, artists participate both formally

and informally in a gift economy, a remnant of older systems of patronage. *Why Are Artists Poor?: The Exceptional Economy of the* Arts (Amsterdam: Amsterdam University Press, 2002).

22. See Alex Kotlowitz on the social isolation of public-housing residents on Chicago's West Side, *There Are No Children Here* (New York: Anchor, 1992).

23. See Dick Hebdige on style as bricolage, *Subculture: The Meaning of Style* (London: Routledge, 1979).

24. Castells, *Informational City,* pp. 82–89.

25. Ibid., p. 82.

26. Howard Becker, *Art Worlds* (Berkeley: University of California Press, 1982).

27. Thornton makes a similar argument about dance subcultures in *Club Cultures,* pp. 116-162.

28. Steve Albini, "The Problem With Music," in *Commodify Your Dissent,* ed. Thomas Frank and Matt Weiland (New York: Norton, 1997).

29. Becker, *Art Worlds.*

30. Gary Alan Fine, *Everyday Genius: Self-Taught Art and the Culture of Authenticity* (Chicago: University of Chicago Press, 2004).

31. When I met new informants in the field over the years I evolved a strategy of initial conversation. I would ask, "So, are you at the (Art) Institute?" This worked quite well because (a) the answer often was yes, and (b) when it was no, the informant would often volunteer his or her other credentials as an artist or artistic aspirant (or at least lover of the arts).

32. Susan Ethridge Cannon, "Concentration of Study Area Education Institutions," report, Arthur Anderson LLP, 2000.

33. Howard Singerman, *Art Subjects: Making Artists in the American University* (Berkeley: University of California Press, 1999).

34. Molotch, *Where Stuff Comes From,* p. 179.

35. Bernard Gendron, *Between Montmartre and the Mudd Club* (Chicago: University of Chicago Press, 2002).

36. Hoban, *Basquiat.*

37. Walter Benjamin, *The Arcades Project* (Cambridge, MA: Harvard University Press, 1999).

38. Susan Buck-Morss, *The Dialectics of Seeing* (Cambridge, MA: MIT Press, 1997), pp. 185–186; Graeme Gilloch, *Myth and Metropolis: Walter Benjamin and the City* (New York: Polity, 1997), pp. 164–165.

39. Georg Simmel, "The Metropolis and Mental Life," in *Georg Simmel on Individuality and the Social Forms,* ed. Donald Levine (Chicago: University of Chicago Press, 1971), p. 326.

40. Jacobs, *Death and Life of Great American Cities.*

41. Richard Sennett, *Flesh and Stone: The Body and the City in Western Civilization* (New York: Norton, 1994).

42. Stephen Kinzer, "Chicago: That Tony-Nominated Town," *New York Times,* May 28, 2002, Arts section, p. 2.

43. Nick Hornby, *High Fidelity* (New York: Riverhead, 1996).

44. Davis, *City of Quartz.*

45. Neva Chonin, "Liz Phair," *Rolling Stone,* Nov. 13, 1997; Greg Kot, "Chicago Singer Liz Phair Is the Voice of the Now (For Better or for Worse)," *Chicago Tribune Sunday Magazine,* Sept. 25, 1994; Dylan Siegler, "Phair's Rise Gave Women More Industry Validity," *Billboard,* July 4, 1998.

46. Tom Billings, "Playing Phair," *Chicago SubNation* 2 (1994): 12.

47. In Siegler, "Phair's Rise."

48. Kristin Tillotson, "Fan-Phair," *Minneapolis Star Tribune,* March 8, 1994.

49. A Proquest Internet search for Phair's name generates a staggering number of articles in papers around the country during her breakout period, most with annoying headlines punning on her name, as with "All's Phair," "Fan-Phair," "Phair-ing Well," and so on.

50. Kot, "Voice of the Now."

51. For all the attention given to these women, it is debatable just how meaningfully they changed the male-centeredness of mainstream music culture. In 2002, *Rolling Stone* released its reader's poll naming the top 100 records of all time. Only ten of the top 100 showcased a female vocalist, with the first woman making her appearance at number twenty-nine (Madonna, with *Ray of Light*). Madonna in fact accounted for half of the female vocalist representation on the list. Artists not acknowledged by the list include Joni Mitchell, Janis Joplin, Jefferson Airplane, PJ Harvey, the Breeders, Blondie, the Pretenders, Sarah McLachlan, the Fugees, Dianna Ross, Joan Jett, Patti Smith, Tina Turner, Aretha Franklin, Björk, Sonic Youth, and Liz Phair.

52. This has not appeared to inhibit the sales for current megastars like Eminem.

53. Eric Boehlert, "Matador Is a Hip but Unprofitable Label Best Known for Launching Liz Phair," *Rolling Stone* 737:19, June 27, 1996.

54. In Kot, "Voice of the Now."

55. Ivor Hanson, "Rock for Rock's Sake Is No Longer Enough," *New York Times,* Dec. 5, 1999.

56. Tom Wolfe, *The Painted Word* (New York: Bantam, 1975).

Chapter 8

1. Dean McCannell, *The Tourist: A New Theory of the Leisure Class* (New York: Schocken, 1976).

2. Zukin, *Culture of Cities.*

3. Lloyd and Clark, "City as an Entertainment Machine."

4. Ritzer, *McDonaldization of Society.*

5. Zukin, *Culture of Cities,* p. 70.

6. Katherine Newman, *No Shame in My Game: The Working Poor in the Inner City* (New York: Vintage, 1999).

7. Florida, *Rise of the Creative Class*; Ross, *No-Collar.*

8. See Grazian on "nocturnal capital" in *Blue Chicago.*

9. Arlie Russell Hochschild, *The Managed Heart: Commercialization of Human Feeling* (Berkeley: University of California Press, 1983).

10. Grazian, *Blue Chicago*, p. 21.

11. Paul Willis, *Learning to Labor: How Working Class Kids Get Working Class Jobs* (New York: Columbia University Press, 1977).

12. Peter M. Blau and Otis Dudley Duncan, *The American Occupational Structure* (New York: John Wiley, 1967).

13. Neff, Gina, Elizabeth Wissinger, and Sharon Zukin, "'Cool' Jobs in 'Hot' Industries: Fashion Models and New Media Workers as Entrepreneurial Labor." Working paper, Columbia University, 2002.

14. Michael Burawoy, *Manufacturing Consent* (Chicago: University of Chicago Press, 1979), p. 83.

15. Ibid.

16. Ibid, p. 85.

17. Mihaly Csikszentmihalyi, *Flow: The Psychology of Optimal Experience* (New York: HarperCollins, 1990).

18. Erik Olin Wright and Rachel Dwyer, "The American Jobs Machine: Is the New Economy Creating Good Jobs?" *Boston Review*, December 2000.

19. In his study of crack dealers, Philippe Bourgois detects a similar overstatement of income by the purveyors of illegal drugs (as opposed to the legal drugs dispensed in a bar), who share with bar workers the trait of earning unreported income, mostly in cash. Bourgois, *In Search of Respect: Selling Crack in El Barrio* (Cambridge: Cambridge University Press, 1995).

20. Clifford Geertz, *The Interpretation of Cultures* (New York: Basic Books, 1973), p. 440.

21. Ibid., p. 447.

22. Neff et al., TITLE, p. 15.

23. Charlotte A. Schoenborn and Patricia F. Adams, "Alcohol Use Among Adults: United States, 1997–1998," *Advance Data from Vital and Health Statistics* 324 (2001).

24. Victor Turner, *Dramas, Fields, and Metaphors: Symbolic Action in Human Society* (Ithaca, NY: Cornell University Press, 1975).

25. Geertz, *Interpretation of Cultures*, p. 432.

Chapter 9

1. Josh Karp, "Wicker Park's Buzzbait Catching Some Big Fish in Chicago" *E*Prairie*, Jan. 29, 2001.

2. Matthew Jaffe, "Best New Place for Media Companies," *Industry Standard*, January 2001.

3. Joel Kotkin, "Here Comes the Neighborhood."

4. Margaret Littman, "Tech Artists Find Wicker Park, Bucktown Great for Business," *E*Prairie*, Feb. 5, 2001.

5. Felicity Barringer and Alex Kucynski, "Industry Standard Says It Will Cease Publication," *New York Times*, Aug. 17, 2001.

6. Jaffe, "Best New Place," p. 1.

7. Michael Indergaard, "Innovation, Speculation, and Urban Development: The Media Market Brokers of New York City," *Critical Perspectives on Urban Redevelopment* 6 (2001): 107–146.

8. Paulina Borsook, "How the Internet Ruined San Francisco," *Salon*, Oct. 27, 1999; Rebecca Solnit and Susan Schwartzenberg, *Hollow City: The Siege of San Francisco and the Crisis of American Urbanism* (New York: Verso, 2000).

9. Neff, "Game Over"; Indergaard, *Silicon Alley*.

10. Robert D. Atkinson, "Technological Change and Cities," *Cityscape* 3, 3 (1998): 129–170; Manuel Castells, *The Rise of the Network Society* (London: Blackwell, 1996).

11. Daniel Bell, *The Coming of Post-Industrial Society* (New York: Basic Books, 1973).

12. Reich, *Work of Nations*.

13. Clark et al., "Amenities Drive Urban Growth"; Richard Florida, "Competing in the Age of Talent: Quality of Place and the New Economy," report prepared for the R. K. Mellon Foundation, Heinz Endowments, and Sustainable Pittsburgh, 2000; Nevarez, *New Money*.

14. According to Florida, "The bohemian index is based on occupational data from the 1990 Decennial Census Public Use Microdata Samples (1% and 5%). It includes the following occupations: authors, designers, musicians and composers, actors and directors, craft artists, painters, sculptors and artist printmakers, photographers, dancers, and artists, performers and related workers. The index is basically a location quotient that measures the presence of bohemians in a region compared to the total national population." Richard Florida, "Bohemia and Economic Geography," 9.

15. Ibid., p. 3.

16. Joel Kotkin, *The New Geography: How the Digital Revolution is Reshaping the American Landscape* (New York: Random House, 2000). p. 3.

17. Lash and Urry, *Economies of Signs and Space*.

18. Writes Lefebvre: "Social relations of production have a social existence to the extent that they have a spatial existence; they project themselves into space, becoming inscribed there, and in the process producing the space itself" (*Production of Space*, p. 129).

19. Klein, *No Logo*, p. 232.

20. Ibid., p. 4.

21. Florida, *Rise of the Creative Class*.

22. Neil Strauss, "Making a Museum out of Music, Part 2: Seattle's Version Has a Do It Yourself Focus," *New York Times*, June 26, 2000.

23. Klein, *No Logo*, p. 43.

24. Eva Pariser, "Artists' Websites: Declarations of Identity and Presentations of Self," in *Web.Studies: Reviewing Media Studies for the Digital Age*, ed. David Gauntlet (New York: Arnold, 2000), p. 62.

25. Steven Henry Madoff, "Art in Cyberspace: Can It Live Without a Body?" *New York Times*, Jan. 21, 1996.

26. Ray Pride, "Plug and Play: Inside the Minds Behind the Museum of Contemporary Art's Version, 2 Digital Media Conference," *New City*, April 18, 2002.

27. Amy Harmon, "Beyond Boosterism: New York's Web Industry May Be Outgrowing Its Image," *New York Times*, Aug. 17, 1998.
28. Piercing Mildred is a pun on James M. Cain's noir classic *Mildred Pierce*. It makes a game out of a favored neighborhood activity, body piercing.
29. Lash and Urry, *Economies of Signs and Space*, p. 27.
30. Lilley and DeFranco, "Geographic and Political Distribution of Arts-Related Jobs in Illinois."
31. Ulrich Beck, *The Brave New World of Work* (Cambridge, UK: Polity Press, 2000), p. 53.
32. Ibid., p. 54.

Chapter 10

1. Peter Marks, "From *Moulin Rouge* to Puccini: Next for Baz Luhrmann, Bringing *La Bohème* to Broadway," *New York Times*, March 14, 2002, p. E1; Ann Midgette, "Loaf of Bread, Jug of Wine, and Puccini Central Park," *New York Times*, June 13, 2002, p. E1; Ann Midgette, "New Stars and Two Debuts in Met's Venerable Bohème," *New York Times*, Oct. 4, 2001, p. E7.
2. Wendy Griswold, *Renaissance Revivals* (Chicago: University of Chicago Press, 1996).
3. Wetzsteon, *Republic of Dreams*, p. 9.
4. Jonathan Larson, *Rent* (New York: Rob Weisbach, 1997).
5. Michael Herzfeld, *Cultural Intimacy: Social Poetics in the Nation-State* (New York: Routledge, 1997), p. 109.
6. Hannerz, *Cultural Complexity*.
7. See Hebdige, *Subculture*.
8. Mark Gottdiener, "Hegemony and Mass Culture: A Semiotic Approach," *American Journal of Sociology* 80 (1985): 1319–1341.
9. Vicki Smith, "New Forms of Work Organization," 7.
10. Beck, *Brave New World of Work*, p. 85–86.
11. Bell, *Cultural Contradictions of Capitalism*, p. 14.
12. Bell, *Coming of Post-industrial Society*.
13. Antonio Negri, *The Politics of Subversion* (Cambridge, England: Polity Press, 2001), cited in Tereza Terranova, "Free Labor: Producing Culture for the Digital Economy," *Social Text* 18:2 (2000): 33-57.
14. François Truffaut, *Hitchcock* (New York: Touchstone, 1967), p. 98.
15. Andres Duany, Elizabeth Plater-Zyberk and Jeff Speck, *Suburban Nation: The Rise of Sprawl and the Decline of the American Dream* (New York: North Point, 2000), p. 14.
16. The critically acclaimed *American Beauty* may be the appropriate 1990s analogue to *Taxi Driver*, in which suburban life is treated as unbearably soul-destroying. While Travis Bickle decries the garbage on the city streets, in *American Beauty* a piece of garbage floating in the wind in the otherwise oversanitized suburban environment is regarded by a sensitive teenage protagonist as "the most beautiful thing in the world."

REFERENCES

Abbing, Hans. 2002. *Why Are Artists Poor?: The Exceptional Economy of the Arts.* Amsterdam: Amsterdam University Press.

Abbott, Andrew. 1999. *Department and Discipline: Chicago Sociology at One Hundred.* Chicago: University of Chicago Press.

___. 2002. "Los Angeles and the Chicago School: A Comment on Michael Dear." *City and Community* 1, 1: 33–38.

Abbott, Carl. 1996. "Thinking About Cities: The Central Traditions in U.S. Urban History." *Journal of Urban History* 22, 6: 687–701.

Abu-Lughod, Janet Lipmann. 1999. *New York, Chicago, Los Angeles: America's Global Cities.* Minneapolis: University of Minnesota Press.

Addams, Jane. 1912. *Twenty Years at Hull House.* New York: McMillan.

Adorno, Theodor W., and Max Horkheimer. [1944] 1994. *Dialectic of Enlightenment.* New York: Continuum.

Aglietta, Michel. 1979. *A Theory of Capitalist Regulation: The U.S. Experience.* New York: Verso.

Albini, Steve. 1996. "The Problem With Music." In *Commodify Your Dissent,* ed. Thomas Frank and Matt Weiland. New York: Norton.

Algren, Nelson. 1942. *Never Come Morning.* New York: Seven Stories.

___. 1947. "The Devil Came Down Division Street." In *The Neon Wilderness.* New York: Doubleday.

___. 1949. *The Man With the Golden Arm.* New York: Seven Stories.

___. 1951. *Chicago: City on the Make.* Chicago: University of Chicago Press.

Allen, Irving. 1984. "The Ideology of Dense Neighborhood Redevelopment." In *Gentrification, Displacement, and Neighborhood Revitalization,* ed. John L. Palen and Bruce London. Albany: State University of New York Press.

Anderson, Elijah. 1990. *Streetwise: Race, Class, and Change in an Urban Community.* Chicago: University of Chicago Press.

___. 1999. *Code of the Street: Decency, Violence, and the Moral Life of the Inner City.* New York: Norton.

Anderson, Perry. 1998. *The Origins of Postmodernity.* London: Verso.

Atkinson, Robert D. 1998. "Technological Change and Cities." *Cityscape* 3, 3: 129–170.

Azerrad, Michael. 2001. *Our Band Could Be Your Life.* New York: Back Bay Books.

Balzac, Honore de. [1843] 1976. *Lost Illusions.* New York: Viking.

Barnet, Andrea. 2004. *All-Night Party: The Women of Bohemian Greenwich Village and Harlem.* Chapel Hill, NC: Algonquin.

Baudelaire, Charles. [1857] 1982. *Les Fleurs du Mal.* Translated by Richard Howard. Boston: David R. Godine.

Baudrillard, Jean. 1981. *For a Critique of the Political Economy of the Sign.* New York: Telos.
———. 1989. *America.* New York: Verso.
Bauman, Zygmunt. 2002. "A Sociological Theory of Postmodernity." In (Eds.), *Contemporary Sociological Theory,* ed., Craig Calhoun, Joseph Gerteis, James Moody, Steven Pfaff, and Indermohan Virk. Oxford: Blackwell.
Beauregard, Robert. 2003a. "City of Superlatives." *City and Community* 2, 3: 183–199.
———. 2003b. *Voices of Decline: The Postwar Fate of U.S. Cities.* New York: Routledge.
Beauvoir, Simone de. 1998. *Beloved Chicago Man: Letters to Nelson Algren 1947–64* London: Gollancz, 1998.
Beck, Ulrich. 2000. *Brave New World of Work.* Cambridge, UK: Polity.
Becker, George. 1978. *The Mad Genius Controversy.* Beverly Hills, CA: Sage.
Becker, Howard. 1982. *Art Worlds.* Berkeley: University of California Press.
Bell, Daniel. 1960. *The End of Ideology.* Cambridge, MA: Harvard University Press.
———. 1973. *The Coming of Post-Industrial Society.* New York: Basic Books.
———. 1976. *The Cultural Contradictions of Capitalism.* New York: Basic Books.
———. 1988. "New Afterword" to *The End of Ideology.* Cambridge, MA: Harvard University Press.
Benjamin, Walter. [1939] 1999. "Paris, Capital of the Nineteenth Century." In *The Arcades Project.* Cambridge, MA: Harvard University Press.
Bensman, Joseph, and Arthur J. Vidich. 1987. *American Society: The Welfare State and Beyond.* South Hadley, MA: Bergin and Garvey.
Berlant, Lauren. 1997. *The Queen of America Goes to Washington City.* Durham, NC: Duke University Press.
Berman, Marshall. 1982. *All That Is Solid Melts Into Air.* New York: Penguin.
Bielby, William T., and Denise D. Bielby. 1999. "Organizational Mediation of Project-Based Labor Markets: Talent Agencies and the Careers of Screenwriters." *American Sociological Review* 64: 64–85.
Blau, Peter M., and Otis Dudley Duncan. 1967. *The American Occupational Structure.* New York: John Wiley.
Bloom, Allan. 1987. *The Closing of the American Mind.* New York: Simon and Schuster.
Bodvarsson, O. B., and W. A. Gibson. 1994. "Gratuities and Customer Appraisal of Service: Evidence from Minnesota Restaurants." *Journal of Socio-Economics* 23: 287–302.
Boehlert, Eric. 1993. "Chicago: Cutting Edge's New Capital." *Billboard,* August.
———. 1996. "Matador Is a Hip but Unprofitable Label Best Known for Launching Liz Phair." *Rolling Stone* 737: 19.
Boon, Marcus. 2002. *The Road of Excess: A History of Writers on Drugs.* Cambridge, MA.: Harvard University Press.
Borsook, Paulina. 1999. "How the Internet Ruined San Francisco." *Salon,* Oct. 27.
Bourdieu, Pierre. 1977. *Outline of a Theory of Practice.* Cambridge: Cambridge University Press.
———. 1984. *Distinction.* Cambridge, MA: Harvard University Press.
———. 1990. *In Other Words.* London: Polity.
———. 1992. *The Rules of Art.* Palo Alto, CA: Stanford University Press.

___. 1993. *The Field of Cultural Production*. New York: Columbia University Press.

Bourgois, Philippe. 1995. *In Search of Respect: Selling Crack in El Barrio*. Cambridge: Cambridge University Press.

Boyer, Robert. 1990. *The Regulation School: A Critical Introduction*. New York: Columbia University Press.

Brenner, Neil. 2000. "The Urban Question as a Scale Question: Reflections on Henri Lefebvre, Urban Theory, and the Politics of Scale." *International Journal of Urban and Regional Research* 24: 2.

Brodsky, M. 1994. "Labor Market Flexibility." *Monthly Labor Review* 117, 11: 53–60.

Brooks, David. 2001. *BoBos in Paradise*. New York: Simon and Schuster.

Buck-Morss, Susan. 1997. *The Dialectics of Seeing*. Cambridge, MA: MIT Press.

Burawoy, Michael. 1979. *Manufacturing Consent*. Chicago: University of Chicago Press.

___. 1991. *Ethnography Unbound*. Berkeley: University of California Press.

___. 2000. *Global Ethnography*. Berkeley: University of California Press.

Bürger, Peter. 1992. "The Institution of Art as a Category in the Sociology of Literature." In *The Institutions of Art*, ed. Peter Bürger and Christa Bürger. Lincoln: University of Nebraska Press.

Burroughs, William S. 1953. *Junky*. New York: Ace.

___. 1959. *Naked Lunch*. Paris: Olympia.

Carroll, Jim. 1978. *The Basketball Diaries*. Bolinas, CA: Tombouctou.

Castells, Manuel. 1972. *The Urban Question*. Cambridge, MA: MIT Press.

___. 1989. *The Informational City*. Oxford: Blackwell.

___. 1996. *The Rise of the Network Society*. Oxford: Blackwell.

Caves, Richard. 2000. *Creative Industries: Contracts Between Art and Commerce*. Cambridge, MA: Harvard University Press.

Chandler, Tertius, and Gerald Fox. 1974. *3000 Years of Urban Growth*. New York: Academic Press.

Chonin, Neva. 1997. "Liz Phair," *Rolling Stone,* Nov. 13.

Clark, T. J. 1973. *The Absolute Bourgeois*. Berkeley: University of California Press.

___. 1984. *The Painting of Modern Life*. Princeton, NJ: Princeton University Press.

Clark, Terry Nichols. 2000. "Trees and Real Violins: Building Post-Industrial Chicago." Working paper, University of Chicago.

___. 2004. *The City as an Entertainment Machine*. Oxford: Elsevier.

Clark, Terry Nichols, and Lorna Ferguson. 1983. *City Money*. New York: Columbia University Press.

Clark, Terry Nichols, Richard Lloyd, Kenneth Wong, and Pushpam Jain. 2002. "Amenities Drive Urban Growth." *Journal of Urban Affairs* 24, 5: 517–532.

Clark, Terry Nichols, and Michael Rempel. 1997. *Citizen Politics in Postindustrial Societies*. Boulder, CO: Westview Press.

Cohen, Adam, and Elizabeth Taylor. 2000. *American Pharaoh: Mayor Richard J. Daley: His Battle for Chicago and the Nation*. New York: Little, Brown.

Collier, Paul, and David Dollar. 2002. *Globalization, Growth, and Poverty*. New York: World Bank.

Coupland, Douglas. 1992. *Generation X: Tales for an Accelerated Culture.* New York: St. Martin's Press.

Crow, Thomas. 1996. *Modern Art and the Common Culture.* New Haven, CT: Yale University Press.

Csikszentmihalyi, Mihaly. 1990. *Flow: The Psychology of Optimal Experience.* New York: HarperCollins.

Davis, Mike. 1990. *City of Quartz.* New York: Verso.

Dear, Michael. 2000. *The Postmodern Urban Condition.* Oxford: Blackwell.

___. 2002a. *From Chicago to LA: Making Sense of Urban Theory.* Thousand Oaks, CA: Sage.

___. 2002b. "Los Angeles and the Chicago School." *City and Community* 1, 1: 5–32.

___. 2003. "Superlative Urbanisms: The Necessity of Rhetoric in Social Theory." *City and Community* 2, 3: 201–216.

Deutsche, Rosalyn. 1996. *Evictions: Art and Spatial Politics.* Cambridge, MA: MIT Press.

Deutsche, Rosalyn, and C. G. Ryan. 1984. "The Fine Art of Gentrification." *October* 31: 91–111.

Dreiser, Theodore. 1957. *Sister Carrie.* New York: Sagamore.

Drew, Bettina. 1989. *Nelson Algren: A Life on the Wild Side.* Austin: University of Texas Press.

Duany, Andres, Elizabeth Plater-Zyberk, and Jeff Speck. 2000. *Suburban Nation: The Rise of Sprawl and the Decline of the American Dream.* New York: North Point.

Du Gay, Paul. 1997. "Organizing Identity: Making up People at Work." In *Production of Culture/ Cultures of Production,* ed. Paul du Gay. London: Sage.

Duneier, Mitchell. 1992. *Slim's Table.* Chicago: University of Chicago Press.

___. 1999. *Sidewalk.* New York: Farrar, Strauss, and Giroux.

Dunn, Jancee. 1994. "Liz Phair," *Rolling Stone,* Oct. 6.

Durkheim, Emile. [1901] 1982. *The Rules of the Sociological Method.* New York: Free Press.

Eastman, Max. 1970. *Art and the Life of Action.* New York: Books for Libraries.

Easton, Malcolm. 1990. "Literary Beginnings." In *On Bohemia,* ed. Cesar Grana and Marigold Grana. New Brunswick, NJ: Transaction.

Eco, Umberto. 1986. *Travels in Hyperreality.* New York: Harcourt Brace Jovanovich.

Edwards, Gavin. 2002. "The Readers' 100 Greatest Albums of All Time." *Rolling Stone,* Oct. 17.

Eeckhout, Bart. 2001. "The Disneyfication of Times Square: Back to the Future?" *Critical Perspectives on Urban Redevelopment* 6: 379–428.

Ehrenreich, Barbara. 1990. *Fear of Falling: The Inner Life of the Middle Class.* New York: Perennial.

___. 2001. *Nickel and Dimed.* New York: Owl.

Engels, Friedrich. 1892. *The Condition of the Working-Class in England in 1844.* London: Swan Sonnenschein.

Esping-Anderson, G. 1996. "After the Golden Age: Welfare State Dilemmas in a Global Economy." In *Welfare States in Transition: National Adaptations in Global Economies,* ed. G. Esping-Anderson. London: Sage.

Fairbanks, Robert. 2003. "A Theoretical Primer on Space," *Critical Social Work* 3: 131–154.

Featherstone, Mike. 1991. *Consumer Culture and Postmodernism*. London: Sage.

Ferris, William. 1988. *The Grain Traders: The Story of the Chicago Board of Trade*. East Lansing: Michigan State University Press.

Filer, Randall K. 1986. "The 'Starving Artist' — Myth or Reality? Earnings of Artists in the United States." *Journal of Political Economy* 94, 1: 56–75.

Fine, Gary Alan. 1996. *Kitchens: The Culture of Restaurant Work*. Berkeley: University of California Press.

___. 2004. *Everyday Genius: Self-Taught Art and the Culture of Authenticity*. Chicago: University of Chicago Press.

Finnegan, William. 2003. "The Economics of Empire." *Harper's*, May.

Firey, Walter T. 1945. "Culture and Symbolism as Ecological Variables." *American Sociological Review* 10: 140–148.

Fischer, Claude. 1975. "Toward a Subcultural Theory of Urbanism." *American Journal of Sociology* 80.

Fitch, Noel Riley. 1985. *Sylvia Beach and the Lost Generation: A History of Literary Paris in the Twenties and Thirties*. New York: Norton.

Flaubert, Gustave. [1869] 1991. *Sentimental Education*. New York: Viking.

Florida, Richard. 2000. "Competing in the Age of Talent: Quality of Place and the New Economy." Report prepared for the R. K. Mellon Foundation, Heinz Endowments, and Sustainable Pittsburgh.

___. 2002a. "Bohemia and Economic Geography." *Journal of Economic Geography* 2: 55–71.

___. 2002b. *The Rise of the Creative Class*. New York: Basic Books.

Franck, Dan. 2001. *Bohemian Paris*. New York: Grove.

Frank, Robert H., and Philip J. Cook. 1995. *The Winner-Take-All Society*. New York: Penguin.

Frank, Thomas. 1997. *The Conquest of Cool*. Chicago: University of Chicago Press.

___. 2000. *One Market Under God*. New York: Anchor.

Frank, Thomas, and Matt Weiland, eds. 1997. *Commodify Your Dissent*. New York: Norton.

Frey, Bruno S. 2000. *Art and Economics*. New York: Springer-Verlag.

Frisby, David. 2001. *Cityscapes of Modernity*. Cambridge: Polity.

Fuentes, Annette and Barbara Ehrenreich. 1984. *Women in the Global Factory*. Cambridge, MA: South End Press

Gagnier, Regina. 2000. *The Insatiability of Human Wants*. Chicago: University of Chicago Press.

Galbraith, John Kenneth. 1998. *The Affluent Society*. New York: Mariner.

Gamson, Joshua. 1994. *Claims to Fame: Celebrity in Contemporary America*. Berkeley: University of California Press.

Gans, Herbert. 1962. *The Urban Villagers*. New York: Free Press.

___. 1974. *Popular Culture and High Culture*. New York: Basic Books.

___. 1995. "Urbanism and Suburbanism as a Way of Life." In *Metropolis: Center and Symbol of Our Times*, ed. Philip Kasinitz. New York: New York University Press.

Garreau, Joel. 1991. *Edge City: Life on the New Frontier.* New York: Anchor.

Geertz, Clifford. 1973. *The Interpretation of Cultures.* New York: Basic Books.

Gendron, Bernard. 2002. *Between Montmartre and the Mudd Club.* Chicago: University of Chicago Press.

Giddens, Anthony. 1984. *The Constitution of Society: Outline of the Theory of Structuration.* Berkeley: University of California Press.

Gilder, George. 1981. *Wealth and Poverty.* New York: Basic Books.

____. 1990. *Microcosm: The Quantum Revolution in Economics and Technology.* New York: Touchstone.

Gilloch, Graeme. 1997. *Myth and Metropolis: Walter Benjamin and the City.* New York: Polity.

Ginsberg, Allen. 1956. "Howl." In *Howl and Other Poems.* San Francisco: City Light Books.

Gitlin, Todd. 1987. *The Sixties: Years of Hope, Days of Rage.* New York: Bantam.

____. 1995. *The Twilight of Common Dreams.* New York: Metropolitan.

Gladwell, Malcolm. 1997. "The Coolhunt." In *The Consumer Society Reader,* ed. Juliet B. Schor and Douglas B. Holt. New York: New Press, 2002.

Glaeser, Andreas. 2000. *Divided in Unity.* Chicago: University of Chicago Press.

Glaser, Barney, and Anselm Strauss. 1967. *The Discovery of Grounded Theory.* Chicago: Aldine.

Gluck, Mary. 2005. *Popular Bohemia: Modernism and Urban Culture in Nineteenth Century Paris.* Cambridge, MA: Harvard University Press.

Gold, Herbert. 1993. *Bohemia: Where Love, Angst, and Strong Coffee Meet.* New York: Simon and Schuster.

Goldberg, Michael. 1994. "The Goose That Laid the Golden Eggs: Wicker Park Was 'Cutting Edge's Capital' Last Year, but What About Now?" *Insider.*

Gottdiener, Mark. 1985a. "Hegemony and Mass Culture: A Semiotic Approach." *American Journal of Sociology* 80: 1319–1341.

____. 1985b. *The Social Production of Urban Space.* Austin: University of Texas Press.

Gramsci, Antonio. 1971. "Americanism and Fordism." In *Selections from the Prison Notebooks.* New York: International.

Grana, Cesar. 1964. *Bohemian versus Bourgeois.* New York: Basic Books.

____. 1994. "French Impressionism as an Urban Art Form." In *Fact and Symbol: Essays in the Sociology of Art and Literature.* London: Transaction.

Grazian, David. 2003. *Blue Chicago: The Search for Authenticity in Urban Blues Clubs.* Chicago: University of Chicago Press.

Griswold, Wendy. 1996. *Renaissance Revivals.* Chicago: University of Chicago Press

Grossman, James. 1977. "African American Migration to Chicago." In *Ethnic Chicago,* ed. Melvin G. Holli and Peter d'A. Jones. Grand Rapids, MI.: Eerdmans.

Gruen, John. 1966. *The New Bohemia.* New York: Grosset and Dunlap.

Guilbaut, Serge. 1983. *How New York Stole the Idea of Modern Art.* Chicago: University of Chicago Press.

Habermas, Jürgen. 1989. *Lifeworld and System.* Vol. 2 of *The Theory of Communicative Action.* New York: Beacon.

Hannerz, Ulf. 1980. *Exploring the City.* New York: Columbia University Press.

___. 1992. *Cultural Complexity.* New York: Columbia University Press.

Hannigan, John. 1998. *Fantasy City: Pleasure and Profit in the Postmodern Metropolis.* New York: Routledge.

Hardt, Michael, and Antonio Negri. 2000. *Empire.* Cambridge, MA: Harvard University Press.

Harvey, David. 1982. *Limits to Capital.* Chicago: University of Chicago Press.

___. 1989. *The Condition of Postmodernity.* Cambridge, MA: Blackwell.

___. 2001. *Spaces of Capital: Towards a Critical Geography.* New York: Routledge.

___. 2004. *Paris: Capital of Modernity.* New York: Routledge.

Hauser, Phillip, and Leo Schnore. 1965. *The Study of Urbanization.* New York: Wiley.

Hawley, Amos. 1950. *Human Ecology.* New York: Ronald.

Hebdige, Dick. 1979. *Subculture: The Meaning of Style.* London: Routledge.

Hemingway, Ernest. 1964. *A Moveable Feast.* New York: Scribners.

Herzfeld, Michael. 1997. *Cultural Intimacy: Social Poetics in the Nation-State.* New York: Routledge.

Hirsch, Paul. 1972. "Processing Fashions and Fads: An Organization-Set Analysis of the Culture Industry." *American Journal of Sociology* 77: 639–659.

Hirsch, Paul, and Mark Shanley. 1996. "The Rhetoric of Boundaryless: How the Newly Empowered and Fully Networked Managerial Class of Professionals Bought Into and Self-Managed Its Own Marginalization." In *Boundaryless Careers,* ed. Michael Arthur and Denise Rousseau. New York: Oxford University Press.

Hise, Greg. 2002. "Industry and the Landscapes of Social Reform." In *From Chicago to LA,* ed. Michael Dear. Thousand Oaks, CA: Sage.

Hoban, Phoebe. 1998. *Basquiat: A Quick Killing in Art.* New York: Penguin.

Hochschild, Arlie Russell. 1983. *The Managed Heart: Commercialization of Human Feeling.* Berkeley: University of California Press.

Hofstadter, Richard. 1966. *Anti-Intellectualism in American Life.* New York: Random House.

Hogan, Will. 1984. "West Town." In *The Local Community Fact Book of the Chicago Metropolitan Area.* Chicago: Chicago Review Press.

Holden, Greg. 2001. *Literary Chicago: A Book Lover's Tour of the Windy City.* Chicago: Lake Claremont.

Holli, Melvin, and Peter d'A. Jones, eds. 1977. *Ethnic Chicago.* Grand Rapids, MI: Eerdmans.

Hornby, Nick. 1996. *High Fidelity.* New York: Riverhead.

Howard, Alan. 1997. "Labor, History, and Sweatshops in the New Global Economy." In *No Sweat,* ed. Andrew Ross. New York: Verso.

Howe, Neil, and William Strauss. 2000. *Millennials Rising: The Next Great Generation.* New York: Vintage.

Hunter, Albert. 1974. *Symbolic Communities: The Persistence and Change of Chicago's Local Communities.* Chicago: University of Chicago Press.

Indergaard, Michael. 2001. "Innovation, Speculation, and Urban Development: The Media Market Brokers of New York City." *Critical Perspectives on Urban Redevelopment* 6: 107–146.

___. 2004. *Silicon Alley: The Rise and Fall of a New Media District.* New York: Routledge.

Irwin, John. 1977. *Scenes.* London: Sage.

Jackal, Robert. 2003. "What Kind of Order?" *Criminal Justice Ethics* (summer): 54–67

Jackson, Kenneth. 1985. *Crabgrass Frontier: The Suburbanization of the United States.* Cambridge: Oxford University Press.

Jacobs, Jane. 1961. *The Death and Life of Great American Cities.* New York: Basic Books.

Jaffe, Matthew. 2001. "Best New Place for Media Companies." *Industry Standard,* January.

Jameson, Fredric. 1984. "Periodizing the 1960's." In *The Sixties Without Apology,* ed. Sohnya Sayres, Anders Stephanson, Stanley Aronowitz, and Fredric Jameson. Minneapolis: Minnesota University Press.

___. 1991. *Postmodernism: or, the Cultural Logic of Late Capitalism.* Durham, NC: Duke University Press

___. 1998. *The Cultural Turn: Selected Writings on Postmodernism.* New York: Verso.

Jamison, Kay Redfield. 1993. *Touched With Fire: Manic Depressive Illness and the Artistic Temperament.* New York: Free Press.

Janowitz, Tama. 1986. *Slaves of New York.* New York: Washington Square.

J.H.K. 1994. "Guyville Java Jive." *Rolling Stone,* Aug. 11.

Judd, Dennis. 1999. "Constructing the Tourist Bubble." In *The Tourist City,* ed. Dennis Judd and Susan Fainstein. New Haven, CT: Yale University Press.

Kantowicz, Edward. 1977. "Polish Chicago." In *Ethnic Chicago,* ed. Melvin Holli and Peter d'A. Jones. Grand Rapids, MI: Eerdmans.

Kasarda, John. 1985. "Urban Change and Minority Opportunities." In *The New Urban Reality,* ed. P. E. Peterson. Washington, DC: Brookings Institution.

Katznelson, Ira. 1982. *City Trenches.* Chicago: University of Chicago Press.

Keil, Roger. 1994. "Global Sprawl: Urban Form After Fordism?" *Society and Space* 12, 2: 131–136.

Kernaghan, Charles. 1997. "Paying to Lose Our Jobs." In *No Sweat,* ed. Andrew Ross. New York: Verso.

Kerouac, Jack. 1955. *On the Road.* New York: Penguin.

Kincheloe, Samuel. 1929. "The Behavior Sequence of a Dying Church." *Religious Education,* 329–345.

Kitagawa, Evelyn, and Karl Taeuber. 1963. *Local Community Fact Book for Chicago Metropolitan Area.* Chicago: University of Chicago Press.

Klein, Naomi. 1999. *No Logo: Money, Marketing, and the Growing Anti-Corporate Movement.* New York: Picador.

Kornblum, William. 1974. *Blue Collar Community.* Chicago: University of Chicago Press.

Kotkin, Joel. 2000. "Here Comes the Neighborhood." *Inc.com.* August 3.

___. 2000. *The New Geography: How the Digital Revolution Is Reshaping the American Landscape.* New York: Random House.

Kotlowitz, Alex. 1992. *There Are No Children Here.* New York: Anchor.

___ 2002. "False Connections," in *The Consumer Society Reader,* ed. Juliet Schor and Douglas Holt. New York: New Press.

Kracauer, Siegfried. [1927] 1989. "The Mass Ornament." In *Critical Theory and Society*, ed. Stephen Bronner and Douglas Kellner. New York: Routledge.

Krell, Alan. 1996. *Manet*. London: Thomas and Hudson.

Krugman, Paul. 1994. *The Age of Diminished Expectations: U.S. Economic Policy in the 1990's*. Cambridge: MIT Press.

___. 2002. "For Richer." *New York Times Sunday Magazine*, Oct. 20.

Larson, Jonathan. 1997. *Rent*. New York: Rob Weisbach.

Lash, Scott, and John Urry. 1994. *Economies of Signs and Space*. Thousand Oaks, CA: Sage.

Laumann, Edward O., Stephen Ellingson, Jenna Mahay, Anthony Paik, and Yoosik Youm. 2004. *The Sexual Organization of the City*. Chicago: University of Chicago Press.

Lazzarato, Maurizio. 1996. "Immaterial Labor." In *Marxism Beyond Marxism*, ed. Saree Makdisi, Cesare Casarino, and Rebecca E. Karl. London: Routledge.

Lefebvre, Henri. 1974. *The Production of Space*. Cambridge, MA: Blackwell.

Lester, Thomas W. 2000. "Old Economy or New Economy? Economic and Social Change in Chicago's West Town Community Area." Master's thesis. Department of Urban Planning and Policy, University of Illinois–Chicago.

Levitine, George. 1978. *The Dawn of Bohemianism*. University Park: Pennsylvania State University Press.

Lewis, Michael. 2000. *The New New Thing*. New York: Penguin.

Lilley, William, and Laurence DeFranco. 1995. "Geographic and Political Distribution of Arts-Related Jobs in Illinois." Study commissioned by the Illinois Arts Alliance Foundation.

Lipietz, Alain. 1985. *The Enchanted World*. London: Verso.

Lipset, Seymour Martin. 1960. *Political Man: The Social Bases of Politics*. Baltimore, MD: Johns Hopkins University Press.

Lloyd, Richard, and Terry Nichols Clark. 2001. "The City as an Entertainment Machine." *Critical Perspectives on Urban Redevelopment* 6: 359–380.

Logan, John, and Harvey Molotch. 1987. *Urban Fortunes: The Political Economy of Place*. Berkeley: University of California Press.

Lopata, Helena. 1964. "The Function of Voluntary Associations in an Ethnic Community: 'Polonia.'" In *Contributions to Urban Sociology*, ed. Ernest W. Burgess and Donald J. Bogue. Chicago: University of Chicago Press.

Lynn, Michael. 2000. "National Character and Tipping Customs." *International Journal of Hospitality Management* 19: 203–214.

Lyotard, Jean-François. 1979. *The Postmodern Condition*. Minneapolis: University of Minnesota Press.

MacCannell, Dean. 1976. *The Tourist: A New Theory of the Leisure Class*. New York: Schocken.

Madeja, Stanley S. 1998. "The Arts as Commodity." *Culture Work* 2, 3.

Mailer, Norman. 1959. "The White Negro." In *Advertisements for Myself*. New York: Putna.

Marcuse, Herbert. 1964. *One-Dimensional Man*. Boston: Beacon.

Markusen, Ann. 2000. "Targeting Occupations Rather Than Industries in Regional and Community Economic Development." Paper presented at the North American Regional Science Association Meetings, Chicago.

Markusen, Ann, Karen Chapple, Greg Schrock, Daisaku Yamamoto, and Pingkang Yu. 2001. "Hi-Tech and I-Tech: How Metros Rank and Specialize."

Markusen, Ann, and David King. 2003. "The Artistic Dividend: The Arts' Hidden Contribution to Regional Development." Project on Regional and Industrial Economics, Humphrey Institute of Public Affairs; available at www.culturalpolicy. uchicago.edu/pdfs/artistic_dividend.pdf.

Marx, Karl, and Friedrich Engels. 1998. *The Communist Manifesto*. New York: Signet

Massey, Douglas, and Nancy Denton. 1993. *American Apartheid: Segregation and the Making of the Underclass*. Cambridge, MA: Harvard University Press.

McBride, David. 2002. "Death City Radicals: The Counterculture in the New Left in 1960's Los Angeles." In *The New Left Revisited*, ed. John McMillian and Paul Buhle. Philadelphia: Temple University Press.

McCannell, Dean. 1976. *The Tourist: A New Theory of the Leisure Class*. New York: Schocken.

McRobbie, Angela. 1980. "Settling Accounts With Subcultures: A Feminist Critique." *Screen Education* 34: 37–49.

___. 1999. *In the Culture Society: Art, Fashion, and Popular Music*. London: Routledge.

Mele, Christopher. 1994. "The Process of Gentrification in Alphabet City." In *From Urban Village to East Village*, ed. Janet L. Abu-Lughod. Oxford, UK: Blackwell.

___. 2000. *Selling the Lower East Side*. Minneapolis: University of Minnesota Press.

Menger, Pierre Michel. 1999. "Artistic Labor Markets and Careers." *Annual Review of Sociology* 25: 541–574.

Miller, Ross. 1990. *The American Apocalypse: Chicagoans and the Great Fire, 1871–1874*. Chicago: University of Chicago Press.

Mishel, Lawrence, and Jared Bernstein. 1994. *The State of Working America 1994–1995*. Armonk, NY: M.E. Sharpe.

Mitchell, Joseph. 1998. *Up in the Old Hotel*. New York: Vintage.

Mizruchi, Ephraim. 1990. "Bohemia as a Means of Social Regulation." In *On Bohemia*, ed. Cesar Grana and Marigold Grana. New Brunswick, NJ: Transaction.

Moehring, Eugene. 2002. "Growth, Services, and the Political Economy of Gambling in Las Vegas, 1970–2000." In *The Grit Beneath the Glitter: Tales From the Real Las Vegas*, ed. Hal K. Rothman and Mike Davis. Berkeley: University of California Press.

Mollenkopf, John, and Manuel Castells, eds. 1991. *Dual City: Restructuring New York*. New York: Russell Sage Foundation.

Molotch, Harvey. 1996. "LA as Design Product." In *The City: Los Angeles and Urban Theory at the End of the Twentieth Century*, ed. Allen Scott and Edward Soja. Berkeley: University of California Press.

___. 2003. *Where Stuff Comes From*. New York: Routledge.

Molotch, Harvey, William Fraudenburg, and Krista Paulson. 2000. "History Repeats Itself, but How? City Character, Urban Tradition, and the Accomplishment of Place." *American Sociological Review* 65: 791–823.

Morris, Meaghan. 1998. *Too Soon, Too Late: History in Popular Culture*. Bloomington: Indiana University Press.

Murger, Henry. [1848] 1988. *Scenes de la Vie de Boheme*. New York: Schoenhof.

Murray, Charles. 1984. *Losing Ground.* New York: Basic Books.

Naremore, James. 1998. *More Than Night: Film Noir in Its Contexts.* Berkeley: University of California Press.

Nasaw, David. 1993. *Going Out: The Rise and Fall of Public Amusement.* Cambridge, MA: Harvard University Press.

Neff, Gina. 2001. "Risk Relations: The New Uncertainties of Work." *Working USA* 5, 2: 59–68.

___. 2002. "Game Over." *American Prospect* 13: 16.

Neff, Gina, Elizabeth Wissinger, and Sharon Zukin. 2002. "'Cool' Jobs in 'Hot' Industries: Fashion Models and New Media Workers as Entrepreneurial Labor." Working paper, Columbia University.

Nevarez, Leonard. 1999. "Working and Living in the Quality of Life District." *Research in Community Sociology* 9: 185–215.

___. 2003. *New Money, Nice Town.* New York: Routledge

Newman, Katherine. 1988. *Falling from Grace: The Experience of Downward Mobility in the American Middle Class.* New York: Vintage.

___. 1999. *No Shame in My Game: The Working Poor in the Inner City.* New York: Vintage.

Norris, Frank. 1903. *The Pit.* New York: Penguin.

Oldenburg, Ray. 1989. *The Great Good Place.* New York: Paragon.

O'Neil, William. 1978. *The Last Romantic: A Life of Max Eastman.* Oxford: Oxford University Press.

Pacyga, Dominic, and Ellen Skerrett. 1986. *Chicago: City of Neighborhoods.* Chicago: Loyola University Press.

Pariser, Eva. 2000. "Artists' Websites: Declarations of Identity and Presentations of Self." In *Web.Studies: Reviewing Media Studies for the Digital Age,* ed. David Gauntlet. New York: Arnold.

Park, Robert E., Ernest W. Burgess, and Roderick D. McKenzie. 1925. *The City: Suggestions for Investigation of Human Behavior in the Urban Environment.* Chicago: University of Chicago Press.

Peterson, Richard A., and N. Anand. 2002. "How Chaotic Careers Create Orderly Fields." In *Career Creativity: Explorations in the Remaking of Work,* ed. Maury Peiperl et al. New York: Oxford University Press.

Peterson, Richard A., and Roger Kern. 1996. "Changing Highbrow Taste: From Snob to Omnivore." *American Sociological Review* 61: 900-907.

Piore, Michael. 1997. "The Economics of the Sweatshop." In *No Sweat,* ed. Andrew Ross. New York: Verso.

Piore, Michael J., and Charles F. Sabel. 1984. *The Second Industrial Divide.* New York: Basic Books.

Plattner, Stuart. 1996. *High Art Down Low.* Chicago: University of Chicago Press.

Polsky, Ned. 1969. *Hustlers, Beats, and Others.* New York: Anchor.

Poster, Mark. 1990. *The Mode of Information.* Chicago: University of Chicago Press.

Pritchet, V. S. 1990. "Murger's *La Vie de Bohème.*" In *On Bohemia,* ed. Cesar Grana and Marigold Grana. New Brunswick, NJ: Transaction.

Regnery, Henry. 1993. *Creative Chicago*. Evanston, IL: Chicago Historical Bookworks.

Reich, Robert. 1991. *The Work of Nations: Preparing Ourselves for Twenty-First-Century Capitalism*. New York: Knopf.

Rice, William. 2002. "A Piece of the Pie." *Chicago Tribune Magazine,* Jan. 13.

Rich, Frank. 2001. "Times Square: Bring Back New Yorkers — and Sex." *New York Times Magazine,* Nov. 11.

Riesman, David. 1950. *The Lonely Crowd*. New Haven, CT: Yale University Press.

Ritzer, George. 1996. *The McDonaldization of Society*. Thousand Oaks, CA: Pine Forge Press.

Rosaldo, Renato. 1993. *Culture and Truth* Boston: Beacon

Rosenzweig, Roy. 1983. *Eight Hours for What We Will*. New York: Cambridge University Press.

Ross, Andrew. 1989. *No Respect: Intellectuals and Popular Culture*. New York: Routledge.

____. 2001. "The Hedonistic Workplace and the Labor Aristocracy of the Internet." Paper presented at Off the Grid: Urban Ethnography and Radical Politics Conference, New York University.

____. 2003. *No-Collar: The Humane Workplace and Its Hidden Costs*. New York: Basic Books.

Rotella, Carlo. 1998. *October Cities*. Berkeley: University of California Press.

Royko, Mike. 1971. *Boss: Richard J. Daley of Chicago*. New York: Dutton.

Sampson, Robert and Steven Raudenbush. 2003. "Seeing Disorder: Neighborhood Stigma in the Social Construction of 'Broken Windows'" *Social Psychology Quarterly* 67: 319–342

Sanchez, Reymundo. 2000. *My Bloody Life*. Chicago: Chicago Review Press.

Sánchez-Jankowski, Martín. 1991. *Islands in the Street*. Berkeley: University of California Press.

Sassen, Saskia. 1991. "The Informal Economy." In *Dual City: Restructuring New York* ed. John Molienkopf and Manuel Castells. New York: Russell Sage.

____. 1994. *Cities in a World Economy*. Thousand Oaks, CA: Pine Forge Press.

____. 1998. "Cities: Between Global Actors and Local Conditions." *1997 Lefrak Monograph*. College Park, MD: Urban Studies and Planning Program.

____. 2001. *The Global City: New York, London, Tokyo*. Princeton, NJ: Princeton University Press.

____. 2002. "Scales and Spaces: A Reply to Michael Dear." *City and Community* 1, 1: 48–50.

Sassen, Saskia, and Frank Roost. 1999. "The City: Strategic Site for the Global Entertainment Industry." In *The Tourist City,* ed. Dennis Judd and Susan Fainstein. New Haven, CN: Yale University Press.

Schatz, Thomas. 1996. *The Genius of the System: Hollywood Filmmaking in the Studio Era* New York: Owl

Schippers, Mimi. 2002. *Rockin' Out of the Box: Gender Maneuvering in Alternative Hard Rock*. New Brunswick, NJ: Rutgers University Press.

Schoenborn, Charlotte A., and Patricia F. Adams. 2001. "Alcohol Use Among Adults: United States, 1997–1998." *Advance Data From Vital and Health Statistics* 324.

Schoemer, Katie. 1996. "Rockers, Models, and the New Allure of Heroin" *Newsweek,* Aug. 26.

Schor, Juliet. 1992. *The Overworked American: The Unexpected Decline of Leisure*. New York: Basic Books.

Schorske, Carl E. 1981. *Fin-de-Siècle Vienna*. New York: Vintage.

Scott, Allen. 1993. *Technopolis: High Technology Industry and Regional Redevelopment in Southern California*. Berkeley: University of California Press.

Scott, Allen, and Edward Soja. 1996. *The City: Los Angeles and Urban Theory at the End of the Twentieth Century*. Berkeley: University of California Press.

Seigel, Jerrold. 1986. *Bohemian Paris*. Baltimore: Johns Hopkins University Press.

Sennett, Richard. 1994. *Flesh and Stone: The Body and the City in Western Civilization*. New York: Norton.

Siegler, Dylan. 1998. "Phair's Rise Gave Women More Industry Validity." *Billboard*, July 4.

Simmel, Georg. 1971. "The Metropolis and Mental Life." In *Georg Simmel on Individuality and Social Forms,* ed. Donald Levine. Chicago: University of Chicago Press.

Simpson, Charles R. 1981. *SoHo: The Artist in the City*. Chicago: University of Chicago Press.

Sinclair, Upton. 1906. *The Jungle*. New York: Bantam.

Singerman, Howard. 1999. *Art Subjects: Making Artists in the American University*. Berkeley: University of California Press.

Sites, William. 2003. *Remaking New York: Primitive Globalization and the Politics of Urban Community*. Minneapolis: University of Minnesota Press.

Smith, Adam. 1793. *The Wealth of Nations*.

Smith, Alson. 1953. *Chicago's Left Bank*. Chicago: Henry Regnery.

Smith, Neil. 1992. "New City, New Frontier: The Lower East Side as Wild, Wild West." In *Variations on a Theme Park*, ed. Michael Sorkin. New York: Hill and Wang.

____. 1996. *The New Urban Frontier*. New York: Routledge.

Smith, Vicki. 1997. "New Forms of Work Organization." *Annual Review of Sociology* 23: 315–339.

____. 2001. *Crossing the Great Divide: Worker Risk and Opportunity in the New Economy*. Ithaca, NY: Cornell University Press.

Snyderman, George, and William Josephs. 1939. "Bohemia: The Underworld of Art." *Social Forces* 18, 2: 187–199.

Soja, Edward. 1989. *Postmodern Geographies: The Reassertion of Space in Critical Social Theory*. New York: Verso.

____. 1996. *Thirdspace: Journeys to Los Angeles and Other Real-and-Imagined Places*. Cambridge, MA: Blackwell.

____. 2000. *Postmetropolis: Critical Studies of Cities and Regions*. Cambridge, MA: Blackwell.

Solnit, Rebecca, and Susan Schwartzenberg. 2000. *Hollow City: The Siege of San Francisco and the Crisis of American Urbanism*. New York: Verso.

Sorkin, Michael. 1992. *Variations on a Theme Park*. New York: Hill and Wang.

Stansell, Christine. 2000. *American Moderns: Bohemian New York and the Creation of a New Century*. New York: Owl.

Stoner, Madeleine. 2002. "The Globalization of Urban Homelessness." In *From Chicago to L.A.*, ed. Michael Dear. Thousand Oaks, CA: Sage

Storper, Michael. 1989. "The Transition to Flexible Specialization in the US Film Industry: External Economies, the Division of Labor, and the Crossing of Industrial Divides." *Cambridge Journal of Economics* 13: 273–305.

Suttles, Gerald. 1973. *The Social Construction of Community.* Chicago: University of Chicago Press.

___. 1984. "The Cumulative Texture of Local Urban Culture." *American Journal of Sociology* 90: 283–304.

___. 1990. *The Man-Made City.* Chicago: University of Chicago Press.

Terranova, Tiziana. 2000. "Free Labor: Producing Culture for the Digital Economy." *Social Text* 18, 2: 33–57.

Thomas, W. I., and Florian Znaniecki. 1927. *The Polish Peasant in Europe and America.* New York: Knopf.

Thornton, Sarah. 1996. *Club Cultures: Music, Media, and Subcultural Capital.* Hanover, NH: University Press of New England.

Toennies, Ferdinand. 1961. "Gemeinschaft and Gesselschaft." In *Theories of Society,* ed. Talcott Parsons, Edward Shils, Kaspar Naegele, and Jesse R. Pitts. New York: Free Press.

Toepler, Stefan, and Annette Zimmer. 1999. "The Subsidized Muse: Government and the Arts in Western Europe and the United States." *Journal of Cultural Economics* 23, 1–2: 33–49.

Truffaut, François. 1967. *Hitchcock.* New York: Touchstone.

Turner, Victor. 1975. *Dramas, Fields, and Metaphors: Symbolic Action in Human Society.* Ithaca, NY: Cornell University Press.

Urry, John. 1990. *The Tourist Gaze.* London: Sage.

Vance, Carole. 1989. "The War on Culture." *Art in America* 47: 39–43.

Veblen, Thorstein. [1898] 1998. *The Theory of the Leisure Class.* Amherst, NY: Prometheus.

Villanueva, Margaret, Brian Erdman, and Larry Howlett. 2000. "World City/Regional City: Latinos and African Americans in Chicago and St. Louis." Working paper no. 46, Julian Samora Research Institute, Michigan State University.

Wacquaint, Loic J. D. 1993. "America as Social Dystopia." In *The Weight of the World,* ed. Pierre Bourdieu. Palo Alto, CA: Stanford University Press.

___. 2004. *Deadly Symbiosis.* New York: Polity.

Ware, Caroline F. 1935. *Greenwich Village: 1920–1930.* New York: Harper Colophon.

Warren, Roland. 1973. *The Community in America.* New York: Rand McNally.

Watson, Noshua. 2002. "Generation Wrecked." *Fortune,* Oct. 14.

Watson, Steven. 1995. *The Harlem Renaissance.* New York: Pantheon.

Weber, Max. 1958. *The Protestant Ethic and the Spirit of Capitalism.* New York: Macmillan.

Weber, Rachel. 2002. "Extracting Value From the City: Neoliberalism and Urban Redevelopment." *Antipode* 43: 3.

Wetzsteon, Ross. 2002. *Republic of Dreams: Greenwich Village, the American Bohemia 1910–1960.* New York: Simon and Schuster.

Whyte, William H. 1956. *The Organization Man.* New York: Simon and Schuster.

Williams, Orlo. 1990. "The Parisian Prototype." In *On Bohemia*. New Brunswick, NJ: Transaction.

Willis, Paul. 1977. *Learning to Labor: How Working Class Kids Get Working Class Jobs*. New York: Columbia University Press.

Wilson, Elizabeth. 2000. *Bohemians: The Glamorous Outcasts*. New Brunswick, NJ: Rutgers University Press.

Wilson, James Q. 1975. *Thinking About Crime*. New York: Basic Books.

Wilson, Jean Moorcraft. 2000. *Virginia Woolf's London*. New York: Tauris Park.

Wilson, William Julius. 1980. *The Declining Significance of Race*. Chicago: University of Chicago Press.

___. 1987. *The Truly Disadvantaged*. Chicago: University of Chicago Press.

___. 1996. *When Work Disappears*. New York: Knopf.

Wirth, Louis. 1928. *The Ghetto*. Chicago: University of Chicago Press.

___. 1939. "Urbanism as a Way of Life." *American Journal of Sociology* 44, 3-24.

Wolfe, Tom. 1968. *The Electric Acid Kool-Aid Test*. New York: Bantam.

___. 1975. *The Painted Word*. New York: Bantam.

Workplace Resource Center. 1999. "Worker Drug Use and Workplace Policies." Report, Sept. 8.

Wright, Erik Olin. 1985. *Classes*. New York: Verso.

Wright, Erik Olin, and Rachel Dwyer. 2000. "The American Jobs Machine: Is the New Economy Creating Good Jobs?" *Boston Review,* December.

Wu, Chin-Tao. 2002. *Privatising Culture: Corporate Art Intervention Since the 1980s*. New York: Verso.

Zorbaugh, Harvey Warren. 1929. *The Gold Coast and the Slum*. Chicago: University of Chicago Press.

Zukin, Sharon. 1982. *Loft Living: Culture and Capital in Urban Change*. Baltimore, MD: Johns Hopkins University Press.

___. 1991. *Landscapes of Power*. Berkeley: University of California Press.

___. 1995. *The Culture of Cities*. London: Blackwell.

INDEX